DATA SCIENCE AND HUMAN-ENVIRONMENT SYSTEMS

Transformation of the Earth's social and ecological systems is occurring at a rate and magnitude unparalleled in human experience. Data science is a revolutionary new way to understand human-environment relationships at the heart of pressing challenges like climate change and sustainable development. However, data science faces serious shortcomings when it comes to human-environment research. There are challenges with social and environmental data, the methods that manipulate and analyze the information, and the theory underlying the data science itself, as well as significant legal, ethical, and policy concerns. This timely book offers a comprehensive, balanced, and accessible account of the promise and problems of this work in terms of data, methods, theory, and policy. It demonstrates the need for data scientists to work with human-environment scholars to tackle pressing real-world problems, making it ideal for researchers and graduate students in Earth and environmental science, data science, and the environmental social sciences.

STEVEN M. MANSON is a professor in the Department of Geography, Environment, and Society at the University of Minnesota, and a Fellow of the Institute on the Environment. He combines environmental, social, and information science to understand complex human-environment systems. He has won awards from organizations including the US National Aeronautics and Space Administration, the University Consortium for Geographic Information Science, and the Ecological Society of America.

DATA SCIENCE AND HUMAN-ENVIRONMENT SYSTEMS

STEVEN M. MANSON
University of Minnesota

CAMBRIDGE
UNIVERSITY PRESS

Shaftesbury Road, Cambridge CB2 8EA, United Kingdom

One Liberty Plaza, 20th Floor, New York, NY 10006, USA

477 Williamstown Road, Port Melbourne, VIC 3207, Australia

314–321, 3rd Floor, Plot 3, Splendor Forum, Jasola District Centre, New Delhi – 110025, India

103 Penang Road, #05–06/07, Visioncrest Commercial, Singapore 238467

Cambridge University Press is part of Cambridge University Press & Assessment, a department of the University of Cambridge.

We share the University's mission to contribute to society through the pursuit of education, learning and research at the highest international levels of excellence.

www.cambridge.org
Information on this title: www.cambridge.org/9781108486286
DOI: 10.1017/9781108638838

© Steven M. Manson 2023

This publication is in copyright. Subject to statutory exception and to the provisions of relevant collective licensing agreements, no reproduction of any part may take place without the written permission of Cambridge University Press & Assessment.

First published 2023

A catalogue record for this publication is available from the British Library.

ISBN 978-1-108-48628-6 Hardback

Cambridge University Press & Assessment has no responsibility for the persistence or accuracy of URLs for external or third-party internet websites referred to in this publication and does not guarantee that any content on such websites is, or will remain, accurate or appropriate.

Contents

Plates can be found between pages 132 and 133.

Figures

Tables

Preface

The primary rationale for this book is that researchers play an essential part in addressing the many challenges that beset the world. Data science is increasingly important to understanding the human-environment relationships at the heart of such critical issues as climate change, natural hazards, and sustainable development.

A narrower reason (undoubtedly shared by many authors) is that I could not find a book quite like this one. There is excellent work out there, and I cite much of it, but I wanted a volume firmly at the intersection of information, social, and environmental sciences. There is much room to bridge various camps of scholarly work on big data and data science for human-environment systems. Of course, there is the risk of trying to cover too much ground or not offering enough. Hopefully, that is not too much of a problem, but we will see.

We have got our work cut out for us. Data science and cognate fields like big data and artificial intelligence hold great promise for studying human-environment systems. At the same time, human-environment research offers much to data science in terms of sophisticated approaches to compelling real-world problems. Serious shortcomings exist in social and environmental data, some of which we are only beginning to see. There are many ways that we can make our methods better and, at the same time, work on the theories underlying the data science of human-environment systems. We also need to be aware of significant legal, ethical, and policy concerns. However, with luck, we can use the tools of data science and human-environment research to address these concerns.

Once we get past the starry-eyed embrace of data science or its often justified critiques, there is a middle ground or many potential middle grounds in human-environment research. This sounds pollyannaish, but we must collaborate to deal with our collective challenges. It can be hard to remain hopeful in the face of what seems to be one unfolding calamity after another – war, famine, disease, disaster – that are usually expressions of long-running human–environment dynamics.

Nonetheless, we are making many forms of progress, and scholarly inquiry is part of that effort.

Data science is also the locus of much interest and training for next-generation scholars and data professionals who will drive future research. I am privileged to work with bright and passionate students who recognize the need for transdisciplinary collaborations among data scientists, domain experts, and everyone else interested in solving our collective problems. It is increasingly their world, and they need all the help we can give them.

Acknowledgments

A big thanks to my family across all our various branches. This book would not have happened without you! A special thanks to my pandemic roommates, Rachel, Anna, Alex, and Gracie (sort of), for putting up with this process. I would enumerate all of the ways that you supported me, but that is another book. Martha and Jerry, thanks for checking in and for your advice. Geoff is an extraordinarily talented artist who did great things with the figures – thank you! Finally, thanks to Linda, Earl, Martha, Jerry, Ben, Monica, Rachel, Anna, and Alex for participating in Coverpalooza 2022.

Thanks to my friends in Minneapolis and beyond. A special shout-out to Kurt, Mike, David, George, and Rachel for your words of wisdom and support; you helped more than you know. Dylan and Em, thanks for being fellow nerds and talking about big data. Thanks too to Peter, Linda, Anne, David, Bridget, and Steve for the conversations!

Thanks to my University of Minnesota family. The good people at the Minnesota Population Center work tirelessly on the original big data, and I learn much from being part of the spatial team with the intrepid Dave, Jonathan, and Tracy. Thanks to the Office of Research and Graduate Programs crew of Wendy, Jen, Mackenzie, and Molly for listening patiently to my book-related caterwauling. Thanks to folks in the Department of Geography, Environment, and Society, especially Glen for being a smart, funny, and oh-so-capable dude who keeps the ship afloat. Finally, thanks to my decanal crew of Jane, Ann, Ascan, Jo, Melinda, and Howie for their sage advice on the patio.

A huge thank you to Emma, Sarah, Sapphire, Abi, Vignesh and Rowan at Cambridge University Press. You dealt with all of the pandemic issues (mine and yours and everyone else's) with grace, wit, and professionalism. Add an extra helping of thanks to Emma for helping get this book off the ground and Sarah for guiding me through my first manuscript and dealing with so . . . many . . . questions.

Thanks to the National Science Foundation and National Institutes of Health for supporting data-centric work at the National Historical Geographic Information System and International Historical Geographic Information System. I sincerely appreciate the support of the College of Liberal Arts at the University of Minnesota for naming me a Scholar of the College.

Finally, I owe special thanks to Julie Santella and Brittany Krzyzanowski for helping me, including getting me to think about the big picture, especially work outside of my geospatial comfort zone.

1

Data Science and Human-Environment Systems

Transformation of the Earth's social and environmental systems is happening at an incredible pace. The global population has more than doubled over the last five decades, while food and water consumption has tripled and fossil-fuel use quadrupled. Attendant benefits such as longer lifespans and economic growth are increasingly joined by corresponding drawbacks, including mounting socioeconomic inequality, environmental degradation, and climate change. Over the past half-century, interregional differences in population growth rates, unprecedented urbanization, and international migration have led to profound shifts in the spatial distribution of the global population. Economic changes have been dramatic as well. The global per-capita gross domestic product doubled while economic disparities grew in many regions (Rosa et al. 2010).

These socioeconomic shifts have affected a host of natural systems and ecosystem services. Demographic shifts and economic development are distal causes of proximate drivers of environmental change, such as fossil-fuel emissions and land-cover change. These changes affect many natural systems, including land-cover composition, soil and water quality, climate regulation and temperature, and vegetation and animal communities. These environmental dynamics have profound implications for human well-being. Flooding, erosion of coastal areas, and drought already affect human societies in many ways, and these effects will grow sharply in coming decades. These shifts are all facets of interlinked human-environment systems that arise from complex interactions among individuals, society, and the environment (Ehrlich et al. 2012).

Can data science help address human-environment challenges? Scientific and policy bodies have called for more and better data and attendant analyses to support the research needed to meet the impacts of rapid human-environmental change (Millett & Estrin 2012). Socioeconomic, demographic, and other social data that can be closely integrated with Earth systems data are essential to describing the continuously unfolding transformation of human and ecological systems (Holm et al. 2013). Of particular interest is big data, or data sets that are larger and more difficult to handle than those typically used in most fields, and data science, the larger field concerned with big data and analysis.

Data science offers advances in processing and analysis for research and policy development. Special issues in leading journals like *Science* and *Nature* highlight the need for new data and methods to help answer a wide array of questions at the intersection of nature and society (Baraniuk 2011). National scientific bodies such as the US National Academy of Sciences, United Kingdom's Royal Society, European Science Foundation, and Chinese Academy of Sciences have issued high-profile calls to develop and use big data to understand and address scientific and policy challenges stemming from human-environment interactions. We also see the advent of specialized journals, such as *Big Data Earth* and the *International Journal of Digital Earth*, that focus on large human-environment data sets.

Researchers and policymakers see data science's promise and pitfalls for human-environment systems. The move toward analyzing vast new data sets redefines disciplines that range from physics to economics to Earth sciences. These data are gleaned from a host of new sensors, internet activities, and the merging of existing databases. At the same time, some of the initial hype around data science and big data has been tempered by how this work plays out in real-world contexts. The fast growth of some forms of data has highlighted the considerable gaps in other kinds. Humans have studied only a tiny part of the world's oceans or a fraction of the millions of species on the Earth's surface. There are also significant gaps in data on people and society over much of the globe. Human-environment data pose many significant unresolved methodological challenges because they represent complex social and environmental entities and relationships that span multiple organizational, spatial, and temporal levels (Kugler et al. 2015). Data science also faces many unsolved challenges around theory development and myriad policy dimensions. Even as vast databases become more readily accessible and tractable, many problems have yet to be addressed, and much of the promise of big data remains just that – a promise unfulfilled.

1.1 Data Science and Human-Environment Research

There is broad interest in using big data for understanding human-environment interactions and attendant issues – including climate change, natural hazards, ecosystem services, and sustainability. This volume brings together these various research streams while assessing the pros and cons of data science for human-environment scholarship. It draws on various sources but focuses almost exclusively on peer-reviewed research. The goal here is to bridge various camps of scholarly work on big data and data science for human-environment systems. Big data and data science are here to stay; maybe not in their current incarnation, but certainly in some form. Addressing the toughest human-environment issues requires scholars to work together across fields. This list includes (and is not limited to) data scientists, statisticians, and computer scientists; domain scientists working on social, environmental, and natural systems; and scholars in policy and law, and arts and humanities.

The nature of global environmental change and other human-environment topics is one of vast spatial and temporal scales in some ways and the hyperlocal in others. One need only look to action around climate change to see how global social and environmental systems are inextricably linked to individual behavior. These incredible scale shifts mean we deal with a vast range of data, methods, and theories across research domains. Scholars also deal with problems that do not neatly fall along human or environmental lines.

Given these pervasive scale-related problems and the inherent complexity they create, it is not surprising that inter-disciplinary and trans-disciplinary research are both seen as necessary; the problems of global change transcend conventional disciplinary inquiry. Global change is often treated largely as an environmental problem, but the environment is not simply an "independent variable"; indeed, global change is a consequence of social processes. *(Pahl-Wostl et al. 2013, p. 40)*

In simple terms, human-environment research is not the domain of any single research field. Doing this research well requires a deliberate commitment to boundary-crossing and integrated scholarship.

Data science is making deep inroads into many kinds of scholarship on human-environment topics, but the literature is splintered. Some of the most extensive work centers on big data (Section 1.2 dives into definitions of big data), primarily focused on providing wide-ranging and generic overviews. These are often trade books that cite primarily from the gray literature or nonpeer-reviewed blogs and web pages. Increasingly these works include research perspectives as data science has ramped up over the past decade. These resources often have an exuberant bent that is driven by just-so case studies that capture the attention of mass media. This large and general body of work directly (or often indirectly) reflects how big data is big business. Data science is vital to a growing array of economic sectors. This commercial success results from big data and data science, which means they are often couched with an optimistic viewpoint with a mercenary perspective at its core. Much of the early writing on big data was commercial, and the authors were understandably looking to sell their products (Wyly 2014).

Much of the early work in big data and data science relied on nonscholarly and nonpeer-reviewed sources. References to blog posts, web pages, and gray literature abound. Informal and nonpeer-reviewed sites will always be essential venues of information on rapidly emerging issues in technology since more deliberate and careful research and subsequent publications can require years. Apart from not being peer-reviewed, the major drawback of these sites is that they too often disappear. For example, the site www.bigdata-startups.com is cited by dozens of academic papers as a source of crucial information; however, it no longer exists beyond partial and fragmented backups in internet archives. Another example is the work of McKinsey & Company, a management consulting firm. This significant proponent of big data published well-cited work at the now-defunct website www.mckinseyonsociety.com, and its articles only live on as informal copies and references.

Scholarly work in data science and big data has proliferated over the past decade. This work falls into several camps and reflects the rapidity with which data science and big data worked their way into the arenas of science agenda setting, funding, and publication. Academia has always been as prone as any other human endeavor to embrace fads, fashions, and folderal (Dunnette 1966). The rapid embrace of all-things-data is driven in part by fashion, but it is also clear that data science approaches work well for many questions, even when there is room for improvement with others. As explored in later chapters, there are also deeper issues in how scholars can, or should, engage with these approaches. This book speaks to many communities in the hope of helping bring them together around a robust data science of human-environment systems.

Social scientists and humanities scholars have long been interested in nature and human-environment relationships. However, the recent increased visibility of human well-being, climate change, environmental justice, ecological resilience, and sustainability have rapidly expanded social science research on the environment. We are also seeing an increase in digital and environmental humanities, areas with an interest in data science as both a methodology and a subject of critical study. Social science and humanities scholarship comprises a large and growing body of perspectives on big data. The majority of this work critiques big data and its role in specific application areas, such as cities or policing, or from a specific perspective, especially in science and technology studies. There is also scholarship, still in the minority, that offers grounded accounts of the promise and drawbacks of big data for particular scientific and policy domains.

Earth, planetary, ecological, and natural scientists have embraced the study of the Earth as an integrated human-environment system. The physical, chemical, and biological impacts of human activities in the Anthropocene have taken on planetary import (Ruddiman 2013). The transdisciplinary field of Earth-system science focuses on ocean, land, and atmosphere processes, recognizing that changes in the Earth result from complex interactions among these Earth systems and human systems. Ecological, natural, and Earth sciences research with data science tends to center on fairly narrowly defined areas of interest, such as using remote sensing for climate change research or geospatial data to study animal movement. In keeping with environmental scientific publishing in general, this work is usually shared via articles, but a growing number of books, predominantly edited volumes, focus on specific research questions.

Information, data, and computer scientists perform much big data research. Many articles and editorials by these researchers call for greater engagement with domain experts to advance big data. One of the goals of this book is to offer these scholars an overview of significant challenges and opportunities in human-environment research. Information and computer science publications provide a mix of general overviews on the computational aspects of big data or advanced information on specific challenges. Articles and edited volumes also offer case studies within narrowly defined research topics. The vast majority of this work is in keeping with the general publishing model of computer sciences, which tends toward shorter pieces in conference proceedings that may or may not be peer-reviewed.

Debates over the potential and problems of data science can be uneven or narrowly defined. Hidalgo (2014) expresses frustration with these problems in his opinion piece "Saving big data from big mouths," which argues that coverage of big data seems to oscillate between uncritical reports or even hyperbolic odes versus underinformed critiques or jeremiads about the big data strawman. Calls for greater collaboration among fields tend to revolve around linking core fields in data science, especially statistics, computer science, and domain fields in the social and natural sciences, and into the arts and humanities. One common complaint is that data science focuses too often on important yet narrow technical and computational considerations. It gives short shrift to many aspects of substantive domain knowledge. At the same time, domain scholars outside of data science run the risk of ham-handedly using data approaches or caricaturing the entire field based on limited engagement. As we explore later, there are many threads to this conversation. There are fundamental differences among fields and their conceptual and epistemological bases. There are marked disparities in funding and infrastructural support for some kinds of work over others that have far-reaching effects on the kinds of questions being asked and answered by scholars of all stripes.

Communication issues between data scientists and domain scholars are related to the need for better communication between human and environmental researchers. Three decades ago, Stern (1993) called for a *second environmental science* that highlighted the need for environmental science to embrace the human. While there have been positive developments in integration, there is much potential for greater collaboration. As Holm and others put it, various important disciplines, mainly social and human, are too often overlooked or neglected as a science, such as law, architecture, history, literature, communication, sociology, and psychology. These are important disciplines to fully understand Earth systems and human motivation and to guide decision-makers. However, they are not routinely seen as fundamental to giving policy advice. Proponents of interdisciplinary research at times relegate human and social science research to an auxiliary, advisory, and essentially nonscientific status.   *(Holm et al. 2013, p. 26)*

Finally, while the focus on relationships between humans and nature anchors most discussion in this book, it is helpful to recognize that this division can be seen as an arbitrary. People have been looking at human-environment systems for thousands of years (Marsh 1864). At the same time, there is a long-standing body of work in *posthumanism* that questions human-centric explanations and correspondingly rejects the dual construction of nature and culture (Braun 2004). This scholarship rejects the concept that nonhuman beings lack agency and embraces the idea that human and nonhuman beings cocreate many spaces. These spaces range from our stomach microbiome to human relationships with animals to interactions with the Earth.

Posthumanism has critics. It can be seen as perpetuating Eurocentric forms of knowledge, as highlighted by Indigenous critiques of posthumanism that argue that the universalizing claims of ways of knowing and being are themselves problematic (Sundberg 2013). For example, there is an ongoing need for Euro-American scholarship to take more seriously Indigenous knowledge, and how the intellectual labor and activist work of Indigenous scholars and practitioners on the mutual interdependence of humans and the environment illustrates how this division may be illusory (Watts 2013). It is important to bear these issues in mind, even as this book primarily uses a human-environment framing as a helpful shorthand for a complex set of dynamics.

1.2 What Are Big Data?

Data science deals with data, unsurprisingly. Data science has subsumed many aspects of big data as a scholarly endeavor, but it is important to consider data and big data on their own. Most scholarly work relies on data harnessed to various methods and concepts. Most researchers can readily point to the kinds of data they use. The simple notion of data as measures of phenomena that we find interesting (e.g., temperature, population counts, or interviews) suffices for most conversations about data science. However, it is essential to dig a little deeper at times and recognize the long and fraught history of data in science. A note on terminology – we will use big data as a plural noun when speaking of the data as such (e.g., "big data are collected") and as a singular noun when speaking of the larger field of big data (e.g., "big data offers perils and promise").

People have collected data for millennia. People twenty thousand years ago were using *tally sticks*, where they would make notches in pieces of wood or bone to keep track of important things, which presumably came in handy for activities such as trading and keeping inventory of possessions (Mankiewicz 2000). Four thousand years ago, people used calculating devices such as the abacus and stored information in libraries. The rise of modern statistics and record-keeping originated in the 1600s and was codified by the 1800s. In the nineteenth century, people used data in ways nearly indistinguishable from how we employ data, statistics, and modeling today to design descriptive measures and find associations in data (Porter 1986). The scientific meaning of data, which underpins big data, came into being in the 1600s. The term *data* is the Latin plural for *datum*, or "what is given" from the verb *dare*, "to give," but it has a deep, contested, and varied history over the centuries for notions of facts or evidence (Rosenberg 2013). Data are not always simple!

The key to understanding data science is understanding that data are made or captured by an observer. Indeed, some scholars would argue that the term data should be better considered as the term *capta*, from the Latin verb *capere*, meaning "to take" (Checkland & Holwell 2006). This book uses "data" since capta is a technical term for what most people think of as data, but it is helpful to consider what the concept implies for big data. Since observers capture data, this information is biased from initial observations to subsequent data handling, interpretation, and analysis. Statisticians spend much time developing new ways to plumb the nuances of data. Social scientists debate endlessly about how data map onto complicated social phenomena like race or trust. Natural scientists are heavily invested in ensuring their instrumentation and observations are free of systemic bias. The humanities have led the charge against naïve realism, noting that data are not the same as related phenomena, despite how they are often treated as inseparable. Nonetheless, despite best efforts to reduce bias in data, it is inescapable (Section 2.3).

Despite (or perhaps because of) big data being a trendy topic, there is no single commonly shared definition. There is an ongoing scholarly conversation around the origins of big data. Diebold (2012) dives into its definition as one of the earlier users of the term during an academic presentation in 2000. He argues that for the field of econometrics, he is likely one of the originators of *big data* as a term that refers to data sets being too large to be used with existing approaches. However, he uncovers several instances of the term before 2001. Weiss and Indurkhya (1998) use the term repeatedly in their data mining textbook, and researchers with the firm Silicon Graphics used it as early as the mid-1990s. Big data is composed of two common words and associated ideas, so perhaps it is not surprising that there are multiple routes to current usage.

Despite being coined almost two decades ago, the definition of big data remains loose. Critical characteristics for many scholars define big data and data science. Among the most long-lived attributes are the "three v's" of big data: volume, velocity, and variety. We will not belabor these because there is a tremendous amount written on them already, but it helps frame the discussion. The v's of big data trace back to a four-page memo written by Laney (2001) in his role as an analyst for the now-defunct Meta Group. Volume refers to where there is much data, orders of magnitudes larger than is commonly used in most research fields. Velocity describes how data are collected and stored at speed or in real-time. Variety refers to how big data have varying degrees of organization and structure, from well-defined tables to text scraped from the web. Beyond these three basic characteristics, there are ongoing conversations on whether big data should have other v's, such as veracity (accuracy of data) and value (the usefulness of data to answer specific questions (Chen et al. 2014). Dozens of definitions relate to the three v's, additional v's, and other characteristics of big data that start with other letters besides "v."

Volume, or the raw amount of data, is central to any definition of big data and data science. Many fields have large volumes of data. Natural science disciplines, including particle physics, astronomy, and genomics, were early adopters of big data approaches. Genomics and astronomy are home to vast amounts of data. They will grow even more because research projects collect amounts of data that were unthinkable even a few decades ago – on the order of ~25 zettabytes per year. The volume of information generated globally doubles every three years, and this pace is increasing (Henke et al. 2016). Key challenges posed by these data are related to their acquisition, storage, distribution, and analysis. Outside of academia, platforms such as Twitter and Facebook collect and monetize large amounts of data, primarily by developing sophisticated analyses of their users to sell advertising.

A tremendous amount of ink has been dedicated to writing about the size of big data and attendant issues of measuring and defining what "big" means. There is not much value in rehashing those arguments here. Perhaps the easiest way to think about it is that context matters. "Big" is relative to the underlying technology and data format; video files are larger than tweets, but their use matters, such as trying to extract semantic understanding. Bigness tends to revolve around the inability of many existing computing systems or approaches to cope with data and the idea that the amount of data is increasing rapidly, exponentially in some cases. Bigness implies we are always moving toward the horizon and will never get there; in that what is big today, will someday be merely large, or just plain old data.

Most authors are careful to note that the term big is almost meaningless, given how increases in storage, processing speed, and analytical power almost always make the big data of yesterday into the small data of today. There are also debates over whether the bigness of data matters when data science in many fields goes well beyond the engineering and computing challenges that are the focus of so much work in big data (for a more in-depth take, see Chang & Grady 2015). As Jacobs (2009, p. 44) puts it, big data are those "whose size forces us to look beyond the tried-and-true methods that are prevalent at that time." However, this definition is (necessarily) vague in order to apply to many specific problems. No matter the measure, the size of global data holdings is increasing (Figure 1.1).

People have attempted to measure how much information exists. The International Data Corporation is a maker of digital data storage and has attendant biases, but it predicts that the amount of data in the world (termed the Datasphere) will grow from 33 zettabytes in 2018 to 175 by 2025 (Reinsel et al. 2018, p. 3). The same study posits that over 75 percent of the world's population will interact in some way with the data and, by definition, contribute to big data. Global data storage capacity is growing and increasingly moving to digital format. In 1986, 99.2 percent of all storage capacity was in analog forms such as paper volumes, and within two decades, 94 percent of storage capacity was digital (Hilbert & López 2011). Measuring data is an imprecise process and often relies on commercial interests using opaque methods, but it is safe to say there is a lot of data out there (more on data in Chapter 2).

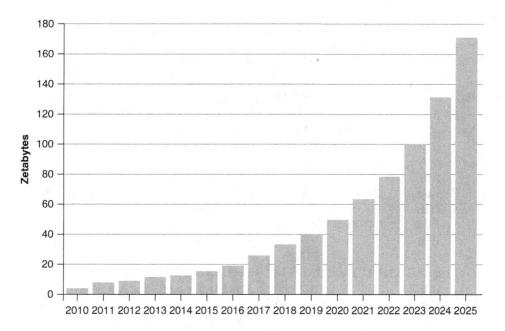

Figure 1.1 Size of global data holdings (2010–25) (Reinsel et al. 2018). Reprinted with permission from the International Data Corporation.

Velocity is another defining characteristic of big data, referring to the rate at which it is collected or moved. Most big data conversations center on how fast data are collected, but velocity also involves how quickly a given computer or processing system can perform calculations over these data. Human-environment data are derived and stored across time frames spanning from paper records and ship logs in the 1600s to real-time digital sensors operating today. The flow rate increases exponentially, especially when considering how scientists use computational modeling to generate simulated data alongside traditional sources (Overpeck et al. 2011). Many data sources are termed *streaming* because they are collected constantly. A significant challenge for human-environment research and data science is developing ways to analyze these data on the fly without assuming that they will be stored in their entirety for later use.

Standard units are used to measure data size. Computer performance has been typically measured by the number of floating-point arithmetic calculations a system can perform in a second (FLOPS). In contrast, data storage is usually measured in bits and bytes. The bit, a contraction of a *binary digit*, is the smallest unit of computer data storage and usually takes the binary value of 0 or 1. A byte is a collection of eight bits and is usually written in binary notation (i.e., 00000000 to 11111111). When using the FLOPS or bytes terminology, we use Greek prefixes to indicate speed or size (Table 1.1). More generically, the suffix *scale* denotes the

Table 1.1　*Size of big data in terms of speed and storage demands*

Prefix	Storage in bytes		Speed in FLOPS		Storage examples
	Byte (B)	1		10^0	Single character
Kilo	Kilobyte (KB)	$1,024^1$	KiloFLOPS	10^3	Half a page of text
Mega	Megabyte (MB)	$1,024^2$	MegaFLOPS	10^6	Photograph
Giga	Gigabyte (GB)	$1,024^3$	GigaFLOPS	10^9	Hour-long video
Tera	Terabyte (TB)	$1,024^4$	TeraFLOPS	10^{12}	One day of Earth Observing System data in 2000 (Frew & Dozier 1997)
Peta	Petabyte (PB)	$1,024^5$	PetaFLOPS	10^{15}	One year of data collected by the United States National Aeronautics and Space Administration in 2015
Exa	Exabyte (EB)	$1,024^6$	ExaFLOPS	10^{18}	One day of data from the Square Kilometer Array (SKA) telescope (Farnes et al. 2018)
Zetta	Zettabyte (ZB)	$1,024^7$	ZettaFLOPS	10^{21}	One year of digital data in 2010 (Gantz & Reinsel 2010)
Yotta	Yottabyte (YB)	$1,024^8$	YottaFLOPS	10^{24}	One day of data generated globally in the mid-2020s (Parhami 2019)

class of computers dealing with data of a given size or processing at a certain speed. For example, a petascale system can perform at least one petaFLOPS calculation or store one petabyte of data.

Variety refers to the enormously heterogeneous nature of data organization, structure, meaning, and sources. Big data can span many knowledge or application domains, especially when dealing with human-environment systems. This range of domains makes these data flexible. They may be collected for one use but be applied to many, even those not foreseen when captured. In some ways, this situation is not new. Land-change scientists, climate modelers, hydrologists, and other researchers on human-environment topics are adept at drawing data from various sources. Traditionally, these data are transformed before analysis, such as converting them into variables for statistical modeling or into layers in a geographic information system. Big data usually focuses on integrating these data on the fly or otherwise reconciling them. It also tackles various sources, from social network posts to remotely sensed imagery to sensor feeds. Much attention on the engineering side of big data is paid to vast data sets and the wide variety of formats and models in place and on the fly.

Definitions of big data center on volume, velocity, and variety. To these three core characteristics, scholars have added others. These include a pile of other "v"s, including variability, veracity, value, and visualization. Some newer characteristics do not begin with the letter "v" because there should be limits on how willing people are to torture the thesaurus. These include relationality, exhaustivity, and complexity. As time goes by, researchers are sure to add more! Human-environment data handily exemplify these features of big data.

- Variability is often added as a fourth "v" that speaks to changes in the first three of volume, velocity, and variety. *Variability* refers to shifts in some characteristic of data, such as volume or variety (Chang & Grady 2015). The ability of a system to accommodate bursts of data or temporary increases in volume, for example, is a significant challenge for computing hardware and software. Variability also poses financial or resource challenges because hardware capable of handling volume or velocity demands is often expensive. Even when computing hardware can be spun down or up quickly to reduce operating costs, their initial capital costs can be high. Their relative performance against each new hardware iteration declines rapidly (Section 3.3.4 goes into different kinds of computing, and their pros and cons for research). Variability also encapsulates the challenges of ingesting, ideally without much human intervention, new data formats, or accommodating a shift in the balance or makeup of different sources.
- *Veracity* refers to data accuracy and other more nuanced notions of data quality. Accuracy and uncertainty have long been topics of interest in science, but they

take on new meaning in the big data era (Gandomi & Haider 2015). One measure of accuracy is how well data match reality, but this entails understanding fitness for use (see Section 2.1). An advantage of big data is that common tasks like correlation or pattern matching can accommodate a fair amount of noise or error in data. Data science practitioners regularly lose hundreds of thousands of data points due to handling and errors, but these losses are typically immaterial when hundreds of millions are in play. At the same time, all data can suffer from systemic temporal, spatial, or attribute biases. Big data can overcome some of these through the sheer number of observations, but not all, and indeed, some kinds of biases are more pronounced with big data. A small but growing body of work examines how bias in data or methods impairs decision-making about people or the environment. Section 2.3 examines bias in data while much of Chapter 4 explores challenges in theory and explanation, and Chapter 5 examines the many real-world challenges of insufficient data.

- *Value* refers to the usefulness of data to answer specific questions and is therefore tied to veracity and accuracy, which often only makes sense regarding fitness for use. Some authors focus primarily on its business value (Ishwarappa & Anuradha 2015), but the value of big data goes far beyond this, applying to a wide array of fields. As data science expands its reach, more disciplines will examine the value of big data for their research questions.

- *Visualization*, or data display, is essential to some data science researchers. They argue that the sheer size of big data means that humans are unusually reliant on visualization as an approach to understanding large and complex data sets. In addition, one of the original computational demands for big data was visualization. This focus has only gained more urgency with the advent of three-dimensional modeling and graphics (Liu et al. 2016).

- *Relationality* is helpful for linking data. Big data become truly big when disparate data sets come together, and relationality refers to data containing common fields that enable two or more data sets to be joined. Name, identifying number, and location are typical examples, but these fields have few limits (boyd & Crawford 2012). Therefore, relational data must be *indexical*, having attributes that uniquely identify them and can be linked to other data. An extraordinary amount of effort and money is dedicated to making data indexical and relational and linking them. Location can make human-environment data information relational in a sense. Researchers can use spatial coordinates to overlay spatial data layers or spatial traces left by people via phones or animals tracked with locational technology. Relationality becomes especially powerful when links between objects can be traced to form networks or graphs of connectivity (more in Section 2.1 on data basics).

- *Exhaustivity* refers to how big data comprises a population of observations and not a sample (Mayer-Schönberger & Cukier 2013). Big data can approach being exhaustive in how their scope spans the, say, entire set of tweets about an event. Of course, these tweets are just a narrow slice of knowledge about a given phenomenon. Data can seem exhaustive in coverage but be sparse in many attributes (Poorthuis 2018). For example, this occurs when collecting global-scale remotely sensed imagery. These data are an excellent source of human and environmental data and encompass the globe but may be collected infrequently or offer only limited attributes (Section 2.4).
- *Complexity* captures how data are complicated in how they are structured. While variety refers to the range of data, some of which can be more complicated than others, complexity refers to data with internal relationships that can require unique or challenging data structures. An example is data on households. Data on individuals, such as age or income, can be more difficult (but more rewarding) when individuals can be linked to other people in a household. In order to answer some kinds of questions, knowing characteristics about a person, like their age or profession, is less helpful than understanding their partnerships or familiar relationships. Complex data require more effort to link, clean, match, and transform (Katal et al. 2013).

There is a cottage industry in defining new "v"s and other letters, so different forms or flavors of big data can exist simultaneously. Other non-"v" attributes can matter a great deal. For example, in exploratory research, data should exhibit *extensionality* (whether one can easily add or change new fields or attributes) or *scalability* (data can expand in size rapidly and hopefully seamlessly) (after Marz & Warren 2015). There is general agreement about volume anchoring the notion of big data, and some scholars argue for using size as the primary criterion (Baro et al. 2015). However, such definitions are always contingent on the broader sociotechnical landscape. Over the past few decades, the overall trend is that yesterday's big data are tomorrow's small data sets as data infrastructure and computing power grow.

1.3 Data Science, a Science about How We Use Data?

Big data is increasingly associated with *data science*, a field centered on gathering and analyzing data. Indeed, big data is probably best considered subsidiary to data science as the latter evolves and takes on a broad array of tasks. One indicator of the rise of data science relative to big data is how internet searches for both big data and data science grew rapidly after 2010. Data science overtook big data around 2014 and grew while big data peaked. The term data science goes back decades, and there is growing recognition that big data and data science are not as new as many of their proponents and practitioners claim. Several data science journals were founded before the term big data became popular, including the *Journal of Data Science* and the *Data Science Journal*, both founded in 2002. Data science goes back decades, and the field is an increasingly essential focus for data, methods, theory, and policy in human-environment scholarship.

Data science has many antecedents. Its roots extend to various forms of data analysis as proposed by scholars from the 1960s onward (Hoaglin et al. 1984). It is well worth reading Tukey's (1962) "The Future of Data Analysis," a substantial (sixty-seven–page) article in the *Annals of Mathematical Statistics*. Tukey speaks primarily to statisticians in a polemic or provocation designed to spur work beyond mathematical and theoretical statistics. He argues that scholars should recognize how much data analysis goes beyond developing models or sample-to-population inferences because it involves using data to guide broader efforts in observation, experimentation, and analysis (Tukey 1962, p. 2). He also notes the importance of the growing use of computers in allowing scholars untrained in statistics or data analysis to engage in both to answer questions of interest. Naur (1974) similarly talks about the science of data, how computers may be used to understand data, and how researchers must explore their data.

Data science in its modern form started to coalesce in the 1990s. C. F. Jeff Wu is credited with one of the earliest uses of data science in statistics, delivering a talk in 1997 entitled "Statistics = Data Science?" (Chipman & Joseph 2016). The presentation described statistics as the trilogy of data collection, data analysis, and decision-making informed by data. It goes so far as to suggest the name of the field should be changed from "statistics" to "data science," and "statistician" to "data scientist." Wu also advocated making statistical education broader and science-driven in ways that focused on modeling with large data sets, and interaction with scholars in other disciplines (more on education in Section 4.4).

Soon after, Cleveland (2001) wrote "Data Science: An Action Plan for Expanding the Technical Areas of the Field of Statistics" as a manifesto for recognizing that statistics is essentially data science, and that statistics should better embrace all elements of data science. This work describes six areas and levels of effort for university-based statistics departments: multidisciplinary investigations (25 percent); models and methods for data (20 percent); computing with data (15 percent); pedagogy (15 percent); tool evaluation (5 percent); and theory (20 percent). This explicit focus by statisticians to embrace data science was partially driven by a desire to claim territory that they saw as theirs, and a recognition that there was significant work outside of statistics in big data and data science.

Other fields were embracing data-intensive science as a precursor of data science. Data-intensive science is one where data drives the initial analysis, and information and knowledge emerge from the data instead of more hypothesis-driven testing and exploration (Newman et al. 2003). This term predates most mentions of big data and data science. However, it captures much of their essence by focusing on the need for large-scale cyberinfrastructure that can manage vast amounts of data in short time frames (real-time in some instances), and allows researchers to sieve through these data to find insight. When data-intensive science is envisioned for the study of biodiversity, for example, there is a distinct focus on the utility of automated exploratory analysis techniques for discovering interesting patterns. These novel findings feed into the development of ecological hypotheses that are proven or disproven by standard data collection and hypothesis testing approaches (Kelling et al. 2009). Similar work was occurring in geography and regional science in artificial intelligence, data-intensive modeling, and automated exploratory analysis of problems, including spatial inter-action and point-pattern analysis (Openshaw 1992).

Big data and data science share some myths (Jagadish 2015). A central myth of big data is that the primary research challenge lies in developing new computing architectures and approaches. Software and hardware are essential, but the focus on eking out performance gains in storage and processing can eclipse more challenging efforts in many workflows, especially in developing better human interfaces and ways for scholars to use their data to answer questions. Another myth is that data science is equivalent to big data. The two are certainly related, but there are also differences, as discussed throughout this book. A researcher can do data science with any data set, and some big data can ignore much of data science. Finally, the commercial impetus for big data has led to a focus on the raw size or volume of big data that has overshadowed other data characteristics. Relatively small and medium-sized data sets can pose their own knotty challenges that benefit from data science approaches, especially regarding variety and veracity.

Learning from data is data science. This focus on learning from data implies the creation of better forms of reasoning from data alongside an interest in understanding how people learn from data. It also means explicitly studying how the field of data science is unfolding in the era of big data. A learning-centric perspective takes a step back from the hype around big data and proposes that it is primarily the "science of learning from data; it studies the methods involved in the analysis and processing of data and proposes technology to improve methods in an evidence-based manner" (Donoho 2017, p. 763). It is not enough to see data science as a loose conglomeration of statistics, machine learning, and engineering centered on large data sets. A broader vision embraces the myriad scientific aspects of data science, centered on a workflow that begins with the initial creation of platforms for collecting these data through to data science approaches to analyzing these data. It also explicitly seeks to engage with other scholarly domains, including human-environment research (Table 1.2).

Table 1.2 *Data science workflow and examples from human-environment research*

Phase	Description	Human-environment research
Platform creation	The purposeful or inadvertent creation of a way to collect data. Big data are often gathered for one reason and then repurposed for others.	Hydrological sensing stations, remote sensing satellites, social media, and telecommunication networks.
Generation and collection	Collation across sources and platforms, alongside developing infrastructure that supports generation and subsequent aggregation and analysis.	Human-environment data span an enormous range, from proprietary social media data sets and telecommunication logs to water and ground sensors to remotely sensed aerial imagery.
Preparation, representation, and transformation	Raw data are cleaned and organized, including selecting subsets of data, conforming to a standard format, extracting information, stripping out ancillary data, and developing machine-readable metadata.	Feature extraction from remotely sensed imagery, semantic encoding, metadata generation, location coordinates from live streaming global navigation system units, and cleaning data with logical rules or statistical approaches.
Aggregation, integration, and reduction	Data are combined from multiple sources and moved into useable databases and systems. It may also involve data reduction and statistical transformations for subsequent processing.	Human-environment projects create high-quality data by combining data streams, such as on-ground and remote sensors, climate data, satellite remote sensing, and demographic data.
Exploration, analysis, interpretation, modeling, and visualization	Conversion into information for decisions and planning. Big data are a noisy mess. They require a range of analytical approaches.	Large swathes of human-environment research increasingly draw on data science, from a network analysis of farming communities to interpreting remotely sensed imagery to complex flood modeling.
Actions or decision-making	Analysis is used to inform planning, decision-making, and behavior. Much of the hyperbole around big data and data science is about how these approaches enable decision-making or risk high-profile missteps.	Much human-environment research addresses complex problems in various systems defined by interactions among human and natural systems.
Science about data science	Actively exploring how data science is practiced.	Human-environment research is theory-driven and empirically grounded, drawing on a broad array of disciplines, making it an ideal venue for advancing data science.

The book structure loosely follows this data science workflow. Chapter 2 examines data and delves into the first two phases of platform creation and generation and collection. Note that the remainder of Chapter 1 refers to methods that subsequent chapters examine more thoroughly. These include artificial intelligence, computers that can think, and the closely related topic of machine learning, the use of computational algorithms that improve solving a problem over time. Standard machine learning approaches include neural networks, computational analogs to biological brains, and deep learning, which uses advanced neural networks. Common tasks include data mining, searching through data for relationships among variables, and data modeling, or creating computational representations of entities and relationships. Chapter 3 looks at methods including artificial intelligence, machine learning, and others and covers the subsequent phases of the data science workflow from data preparation to visualization. Chapters 4 and 5 are on theory and policy, respectively. They tackle various components of taking action or making decisions with data science and doing science about data science.

1.4 Why Is Data Science Growing?

Like most sea shifts in technology, there are many reasons why big data, data science, and cognate fields are rising in prominence. The fact that these fields are hard to define underscores that they are messy and evolving. It is no surprise that the reasons for their existence and growth are also messy and evolving. Here, we focus on three primary reasons: computation is becoming less expensive, networking is becoming near-ubiquitous in many settings, and big data is big business. There are other reasons why data science and data are garnering so much attention. However, these three are an excellent starting place to explore data science for human-environment systems.

Computation is getting less expensive, or seen another way, more powerful for a given unit of cost. The potential of big data relies on continued evolution in data storage, memory handling, and computational power. This growth focuses on basic technology, such as more powerful computer chips, better storage, and faster networking. It also involves new technologies, such as distributing computing over many different machines or harnessing specialized processing units to compete with supercomputers (more in Section 3.3). In many ways, the story of data science is part of an older tale of *Moore's law* (computing power doubles in power on a regular schedule) and the attendant drop in the cost of computing power.

Increased computational power leads to more data. One route is the ease of generation afforded by computing, such as when a person can take a few seconds to send out a picture or message via social media. Another is how computers are increasingly the creators of information. The *Internet of Things (IoT)* refers to networked, low-cost, and ubiquitous computing in everyday devices (more in Section 3.3.5). Computation is inexpensive to the point where devices as prosaic as toasters and alarm clocks are networked. They combine computer processing with near field communications (e.g., Bluetooth or radio-frequency identification), real-time location sharing (via satellite systems or triangulating access to cell phone towers or wireless access points), and the use of embedded sensors for motion and other physical characteristics (Ashton 2009).

Computation is increasingly embedded in everyday higher-end devices such as personal automobiles. They go beyond using data to help run the vehicle and become tools for collecting vast amounts of data about the environment that can be monetized and shared (more on mobile data collection in Section 2.2.5). Some estimates contend that the IoT generates more than half of the world's internet traffic (Networking and Information Technology Research and Development [NITRD] 2016). The world reached a tipping point in 2007 when sensors and systems generated more data than could be kept in the entirety of global storage (Baraniuk 2011). Computerization is also at the heart of automated content generation ranging from search results and news stories generated with no human oversight to procedurally derived landscape simulations and computerized sensors collecting and sharing information without pause.

Second, advances in computation are enhanced by the promulgation of networks. The Internet and other networks offer more one-to-one and one-to-many forms of communication, such as texting a friend or making an environmental sensor reading available to many users. The move to cloud computing for many universities and researchers exemplifies this move to networks. While much data are held on private clouds or other forms of enterprise storage, growing amounts are stored on the public cloud and analyzed there instead of being downloaded first and then examined (Section 3.3.4). Much more data is being generated by converting existing analog data (e.g., scanning paper books or documents to make digital copies) or from the vast array of born-digital media, including text, photographs, and video (Kugler et al. 2017). These are usually shared or promulgated across networks, inducing even greater demand. Dozens of human-environment domains embrace data science as a mode of inquiry and rely on networks for data and analysis. They deal with petabytes of data ranging from human epidemiology and mobility analysis to atmospheric dynamics, radar-based remote sensing, and climate modeling (Fox & Chang 2018).

Computing and networking combine to make data pervasive. Data permeate every facet of the human sphere and are rapidly impinging on the natural world. Throughout this volume, we will explore the data collection that is encompassing the globe and how these data and attendant analyses affect many human-environment systems. Another consideration is that network information itself is also a new form of data, particularly how researchers can examine relationships in a social network. Greenfield (2010) terms this move toward ubiquitous computing *everyware* (a portmanteau of *every*where and soft*ware*), in that software (and by definition, the underlying computing infrastructure) are enmeshed into many facets of daily life. In essence, computation is becoming pervasive in how it is being embedded in a broad range of objects and services. It is ubiquitous in how it is found in many locations, especially as people around the globe adopt mobile telecommunications, and in how technology becomes increasingly locationally aware (Kitchin & Dodge 2011).

Third, big data is big business, pure and simple, leading to many issues in using data science approaches for research. While estimates vary, the value of big data is likely in the trillions of dollars, including using consumer locational data (US $600 billion), gains in the public sector in Europe (US $300 billion), or US health care system (US $300 billion) (Manyika et al. 2011). Even if only partially correct and increasingly outdated, these estimates are staggeringly large and illustrate how any seemingly valueless single datum, when joined to others, can be made into big business. Recent estimates of specific industries, such as artificial intelligence (US $330 billion) or data analytics (US $60 billion), point to the importance of data science and big data across sectors (Holst 2021). Section 5.3 looks at open data and examines how many policymakers see data science and big data as increasingly essential to national security and growth. For example, China is predicted to have the largest datasphere globally by 2025, in large part because it is building a vast video surveillance system (Reinsel et al. 2018). The desire of governments and the private sector worldwide to spy on people drives the growth in data and new ways of analyzing them.

Critically, data science has a large footprint outside of academia, and it behooves scholars to understand how this work affects their research. The rampant growth in data and surge in data science owes much to largely unmonitored data collection tied to advertising dollars. These advertising dollars support the optimistic and opportunistic growth in data science hardware and software through an enormous computing ecosystem (Wyly 2014). Writ large, big data and data science are integral to *surveillance capitalism*, a socioeconomic system driven by the collection and commodification of data (Zuboff 2019). Researchers have an ethical imperative to understand how data science's underlying business orientation can lead to unscientific or harmful inquiry. Section 5.1 visits these and other ethical issues that data science poses for researchers and others.

1.5 Data Philosophy

Any broader discussion of data science should include contributions from the philosophy of science. This field has been a freestanding discipline for decades and draws on centuries of antecedent scholarship. In particular, the subfield of science and technology studies poses important questions for data science and big data. Feenberg (2009) argues that the philosophy of technology is separate from the philosophy of science, given that the focus on truth in science is different from the focus on control of technology. Others would contend that both fields are part of the larger corpus of work that examines science and technology's epistemological and ontological underpinnings, alongside contextualizing their social and political nature (expanded in later work; Feenberg 2017). Either way, data science needs data philosophy.

When defining big data or data science, it is vital to go beyond computational issues and examine the theoretical or epistemological stance of the work. In this view, part of understanding data science and big data is discerning the underlying commercial motives of much of this work (Wyly 2014). More broadly, data science cannot be understood outside social, economic, and cultural contexts. There is so much money in big data that commercial dimensions are inescapable. Data science has social, economic, political, and cultural impacts that can only be barely discerned or guessed in some contexts. Kitchin (2014a) draws on science and technology concepts in describing *data assemblage*, where data must be seen as part of a vast, intricate, and largely unplanned enterprise intimately tied to culture, politics, and socioeconomic systems.

Another view of big data describes it as the interplay of technology, analysis, and mythology. The technology and analysis are straightforward. The former refers to how data tax existing computational methods, the latter, the ability to draw on large data sets to develop causal claims. The notion of big data as mythology helps situate its claims to develop new insights "with the aura of truth, objectivity, and accuracy" (boyd & Crawford 2012, p. 662). Others have called this mythology a form of hubris because it relies on blind faith in the power of data science to distill the essence of very complicated systems. There is an "often implicit assumption that big data are a substitute for, rather than a supplement to, traditional data collection and analysis" (Lazer et al. 2014, p. 1203). *Critical data studies* examines how the data, methods, and assumptions of big data are rarely as simple or innocent as made out by its proponents (Iliadis & Russo 2016). It is also invested in establishing an ethical basis for data science and exploring its broader social and environmental ramifications.

Data science also studies the science of data. Research in science and technology studies often employs ethnographies of data science in action. In contrast, data scientists have called for using big data tools to understand how scholars, policymakers, and others work with data. Examples include using metaanalysis and citation studies to illuminate how science works. The work of Ioannidis (2008) exemplifies this trend by looking at vast swathes of medical research and illuminating issues such as the difficulty of replication or apparent misuse of statistics in reporting significance. Data science has a similar potential to help clarify the advantages and disadvantages of different workflows across studies or computing ecosystems. This research encourages conformity in scientific practice and replicability when running experiments and handling data (more on reproducibility in Section 3.4).

We dip into science and technology work on data in a few places in this volume, but it is hard to do it justice. Instead, we can follow the lead of geographer Muki Haklay (2013), who writes on data, technology, and democracy and draws on work in the philosophy of science (Dusek 2006; especially Feenberg 2002). This scholarship applies to spatial data in ways that resonate with big data and data science conversations. Haklay frames technology as the interplay between *values* and *autonomy*, borrowing from and expanding on Feenberg's (2009) formulation (Table 1.3). Technology is either value-neutral or value-laden and either humanly controlled or autonomous.

Under this schema, the term *value* connotes whether a technological means is tied to specific ends or outcomes. If technology is value-laden, it can be judged as falling somewhere along a continuum of good-to-bad for specific circumstances or issues that go beyond the technology itself and affect human or environmental systems. In this view, a method like artificial intelligence is assumed to have inherent characteristics that will lead to good outcomes, like freeing humans from

Table 1.3 *Technology in terms of values versus agency*

Technology is …	Autonomous	Humanly controlled
Value-neutral: Complete separation of means and ends	Determinism: Technology has universal qualities that make it evolve along a set trajectory independent of human intervention	Instrumentalism: Liberal faith in progress in which technology is value-free but the ends or purposes are important
Value-laden: Form a way of life that includes ends	Substantivism: Means and ends are linked in systems via political and economic dynamics	Constructionism: Humans can modify technological trajectory and choose among alternative means-ends systems, although dystopian facets of technology are never far away

(Adapted from Feenberg 2009)

unnecessary labor, or bad outcomes, like leading to a robot uprising against humans. In comparison, seeing technology as value-neutral implies that it may only be judged on its own terms, such as whether one technology is more efficient or efficacious than another. A given computational approach is better or worse solely in the context of the technology itself, such as the speed with which an analysis can be completed or the accuracy with which a given artificial intelligence method can classify data.

Autonomy speaks to the degree of human control over technology. Specifically, it recognizes that while humans create technology, there is disagreement about the extent to which the growth and nature of technology are either undirected or almost preordained in the sense that there is some inherent evolutionary path. Asked another way, how much of a role do humans play in conducting the direction and use of technology? For example, assuming that artificial intelligence is autonomous means that this technology will evolve independently of human intervention. It does not matter where the technology is created or who is developing it – most efforts will end up with similar technologies or outcomes. In contrast, if technology is human-controlled, then artificial intelligence's nature can be shaped, and humans can and must control the ends to which it is applied.

This values and autonomy framework places data science into one of four categories.

- *Instrumentalism* sees data science as a value-neutral tool that humans control. This view characterizes much of the published work by the proponents of big data, data science, and much work about technology. Explicitly or implicitly, this control is seen as being in the service of making the world a better place. However, it is impossible to ignore the potential for adverse outcomes. Chapter 5 explores policy dimensions of data science and how big data is often considered an unalloyed opportunity to develop solutions to many of the world's human and environmental ills. The instrumentalist view recognizes that data can be misused, but the chances of negative impacts are lessened by having a firm and deliberate hand on the tiller. However, this conception often contends that most of the negative impacts can be attenuated and be outweighed by the benefits.
- *Determinism* shares with instrumentalism the assumption of value-neutral technology, but technology has imperatives independent of its larger social context. Regardless of society's organizing principles, data science will be energetically used to pursue the goals of that system and inherit its values. Technology is simply an extension of the overarching socioeconomic system while retaining much autonomy. We discuss this later in the context of informational sovereignty for firms and nations where data are seen as integral to the ability of nations and firms to use data science to further their interests.

- *Substantivism* agrees with determinism that technology is autonomous but differs in seeing it as having inbuilt values. Substantivism focuses on the adverse effects of power, control, and domination. In this view, data science and big data have an inbuilt tendency toward centralizing power and control. Surveillance exemplifies the substantivist argument that data mining or artificial intelligence technologies are essentially autonomous and tend to support the centralization of power. Technology channels social norms but is not overly guided by them. Societies will choose among rationales to support surveillance – security, safety, commerce – that fit within their guiding ethos, but they will ultimately choose surveillance. Section 5.1.2 explores the control of data and how data mining and artificial intelligence are used to advance a surveillance society where people and groups are tracked.
- *Constructionism* is the longstanding body of thought that posits that our shared understanding of many facets of the world, including technology, are developed via social processes, including communication and power dynamics. Constructionism posits that a range of social imperatives shapes technology and, in turn, will shape many aspects of society (Feenberg 2002). This view shares with substantivism the notion that technology is not value-free, but it tends to be more optimistic than substantivism at the prospect of human control. Constructionism is not always sufficiently attuned to how socially contested technology can be, but it offers a sizeable academic arena that accommodates many views (Bantwal Rao et al. 2015). Constructionism is valuable in examining how technology is value-laden and contends that it is crucial to assess the ends to which technology can be applied. One corollary is that data science can offer myriad outcomes depending on the norms and regulations that govern its use, so humans should direct the growth and use of technology. Of course, it is an open question on how to best guide the development and application of technology. Throughout this volume, we will examine whether and how humans can control the data science of coupled human-environment systems.

The tensions among the different schools of thought on technology around issues of autonomy and value illustrate how important it is to go beyond many of the common critiques of big data and data science. This volume focuses on many practical issues in data science but will occasionally foreground the discussion of values versus autonomy. For example, an instrumentalist approach recognizes that bias is a problem. It argues that biased data can be eradicated by developing better and more objective observing systems. In contrast, a constructionist approach contends that the problem of bias goes deeper than ridding data of errors or modifying data collection systems. Feminist constructionist work in particular demonstrates that data are theory-laden artifacts within a larger context or knowledge infrastructure of people, things, and institutions and tied to webs of race, gender, ethnicity, and class among others human characteristics (D'Ignazio & Klein 2020). There are no unbiased data because observations are inextricably entwined with messy layers of human systems.

These differences among ways of seeing technology can seem arbitrary or abstract, but they say a lot about data science for human-environment systems. They reflect a longstanding engagement in the philosophy of science and cognate fields around the subtlety of data and its technological trappings. For the busy researcher who wants to get on with their work in data science, it can be hard to extract concrete lessons from the long and complicated history of scholarship in the philosophy of science, but these lessons must be heeded. At the most basic, there is a need to recognize the essential concept that data are fundamentally a social construction that defines research of all stripes (Latour 1986). There are times when knowing that data are social artifacts does not matter, but there are times when it does, as explored throughout this volume.

1.6 Promise and Pitfalls of Data Science for Human-Environment Research

Despite the great promise of data science for studying human-environment systems, there are also significant challenges. This book has six chapters: this introduction and a conclusion bookend four chapters that examine the pitfalls and promises of data science for human-environment scholarship in the focal areas of data, methods, theory, and policy. This detailed examination will draw widely on work in the social, natural, and information sciences and double back to examine the promising ways in which big data allows researchers, policymakers, and other stakeholders to understand human-environment dynamics better. Each chapter has a penultimate "focus" subsection that dives into a topic to tie together some of the chapter's main points.

Chapter 2 examines how data science grapples with big human-environment data. These data fundamentally change how many scholars answer a range of human-environment questions, but these data come with new challenges. They have large volumes because they span broad spatial and temporal extents, are collected across multiple scales, and have increasingly high resolutions. These data have high velocity because they are continuously collected and manipulated via a broad array of sensing platforms, ranging from ground-based stations and satellite remote sensing to new sources like social network data. Human-environment data often have complex structures and have many forms of bias, error, or uncertainty. The chapter concludes with a look at the field of remote sensing, which has for decades used many of the tools now considered central to data science, and how remotely sensed imagery has long served as one of the best sources of human-environment information.

Chapter 3 looks at data science's methodological shortcomings and opportunities for manipulating and analyzing big human-environment data sets. Core approaches, including big data, machine learning, or artificial intelligence, are challenged by the complex nature of human-environment data. These data represent various social and biophysical entities and interactions across multiple spatial and temporal scales. Data science can adopt ongoing research in the information sciences on data lifecycles, metadata, ontologies, and data provenance. Many data science challenges are computational, as are the solutions, such as parallel, distributed, cloud, and high-performance computing. Many of these approaches have been developed for generic big data and adapted to spatiotemporal data. Human-environment research also benefits from advances in smart computing, embedded processing and sensing, and the IoT. Another class of methodology central to data science is the burgeoning interest in formalizing and supporting how we share data, workflows, and models in the name of scientific reproducibility. The chapter concludes with a focus on research on handling spatial and temporal patterns and processes and examines this work in the context of global land use and land cover data.

Chapter 4 digs into the many outstanding questions and concerns around the theory of data science and big data. Data science is often positioned as being theory-free or purely inductive, which oversimplifies a more complicated discussion around the various epistemologies of data science. There is an emerging consensus on linking core approaches in data science, like machine learning or modeling, to domain knowledge to overcome some of the challenges of theory development in data science. Work in these substantive areas of inquiry demonstrates the value of linking existing research with smaller data sets to work with larger ones. How these various forms of data science play out in human-environment research is conditioned on the kinds of data science training students receive and how this education relates to competing conceptions of what data science should look like as an academic field or commercial enterprise. Many of the conversations about theory development and data science are happening in the context of the science of cities and the concept of smart cities as coupled human-environment systems.

Chapter 5 concerns the significant policy considerations arising from the profound legal, social, political, and ethical dimensions of data science in general and for human and environmental systems. Data science is touted by many scholars, policymakers, and others as holding extraordinary promise for making policy-making better for people and the planet. Skeptics of the rapid and largely uncontrolled roll-out of data science in many policy arenas have identified a range of harms and argue for policy interventions. There are many policy and data science issues related to discrimination, data dispossession, surveillance, and privacy and consent. How data science plays out in many human-environment contexts is governed in part by data divides, the gaps between people and places in how they may access and use data science tools. Also crucial to successful science and policymaking is open data as defined by a range of competing public and private interests. Dilemmas and potentials of data science for policy development are brought to the fore in debates around the use of data science in sustainable development around the globe.

Chapter 6 wraps up with a discussion of how data science is assuming an increasingly prominent role in examining and addressing human-environment dynamics. Data science is here to stay. It is a valuable way to understand human-environment dynamics and a source of challenges in how the dynamics play out for people and the environment. This final chapter revisits some of the conversations around how data science can, or should, be carried out across research domains. It also delves into issues around the role of data and research in democracy and decision-making. Finally, it brings together many of the threads teased apart in the volume by examining the promise and pitfalls of scientific infrastructure as one of the primary ways forward in applying data science to advancing our knowledge of coupled human-environment systems.

2

Data Gaps and Potential

For all the potential of big data and data science, scholars wanting to untangle human-environment interactions face many gaps in big data. Human-environment data handily exemplify many of the characteristics of big data. They have high volumes, orders of magnitudes more extensive than commonly used in most research fields, resulting from repeated observations over time and space (Jacobs 2009). These spatial and temporal data are often collected and analyzed across multiple scales. They exhibit high velocity, with data being collected and stored in or near real time from an extensive array of sensors at sea, on land, and in air and space, alongside data collected via social networks and internet sources. Human-environment data also exhibit incredible variety in the domains that they relate to and the structures and data models necessary to conduct research. They often represent complex social and biophysical entities and relationships that operate at multiple levels of an organization, over space, and through time. These data also push the boundaries of other characteristics, including value for answering specific questions and veracity in terms of accuracy and fitness for use. The chapter concludes with a dive into remotely sensed imagery as both a research area and data source that captures many of the dynamics around the gaps and potential of data science for understanding human-environment systems.

2.1 Data Basics

Big data for human-environment systems have characteristics that make these data challenging and powerful for research. Data are often divided between quantitative and qualitative and have spatial, temporal, or attribute dimensions. Data have a scale defined by resolution and extent and can be multiscalar. Data have degrees of accuracy with respect to the phenomena being measured and varying degrees of interoperability with other data sets. Depending on the source, data can be unstructured or structured and can be wrangled into various data models that describe how data are formatted, stored, and analyzed. These and other characteristics define how data may be used in human-environment research.

A standard divide is between quantitative and qualitative data, which takes on extra valence when dealing with large data sets. *Quantitative data* are those that may be measured and expressed using numbers. *Qualitative data* are descriptions of properties, attributes, and other identifiers expressed nonnumerically as text, audio/video, interview transcripts, or observation notes. While data science is predominantly focused on quantitative data, the divide between quantitative and qualitative information is narrowing due to data science methods. Data mining and other approaches are getting better at extracting meaning from qualitative data, making it quantitative and machine-legible. Importantly, as discussed later, it is increasingly apparent in many circumstances that qualitative data in and of themselves, and not translated into quantitative data, are essential to many forms of successful analyses.

Human-environment data have characteristics that can be divided among spatial, temporal, and attribute domains.

- These data are often *spatial* because they have a physical location where an object is located, an event has occurred, or a measure was taken. Spatial characteristics relate to location but can be expressed in various ways. Data can outline points, lines, polygons, graphs, and other geometric models. They are usually specified as coordinates on the Earth's surface, such as latitude–longitude or other systems.
- Human-environment data always have a *temporal* component corresponding to when data were collected, modeled, or simulated, representing a period of human or environmental phenomena. This component may be a single instant in time, such as snowpack height at a given point, span a period like total snowfall over a year, or be more complicated, such as the periodicity of snowfall per year over a decade.
- Finally, most data have *attributes* in the form of measurements for characteristics or features of an object or event. These attributes can range from raw measures, such as precipitation or snow depth at a specific measuring station, to modeled characteristics, such as average annual rainfall or snowfall calculated over a decade for a preexisting region like a county or ward.

Most data also have a *scale* defined by resolution and extent. Spatial *resolution* refers to the area or object measured or for which a measurement is made. For census data, it may be a person or a household. However, these data are typically aggregated to some local to regional measures, such as a United States census tract, Canadian dissemination area, or Ethiopian woreda. For remotely sensed imagery, it corresponds to the smallest area for which meaningful spectra can be measured, such as a 100 m^2 plot of ground. These data usually also have a spatial *extent*. Census data are typically collected for a nation (so the extent is that nation's borders). In contrast, remote sensing typically captures a single image of a given size; for example, a Landsat Thematic Mapper image is about 185 km by 172 km. That said, spatiality gets complicated quickly. Remotely sensed images are mosaicked together to create a data set with a global spatial extent. Big spatial data is a trillion-dollar business (Henke et al. 2016, p. 2)

Temporal scale refers to the timing of how observations are made. Analogous to spatial extent, temporal extent refers to the period in which data are collected or the frequency with which a given location is sampled. In the case of aerial vehicles, images are collected at intervals chosen by the producers, while satellites take imagery with a frequency tied to their orbits around the Earth. Temporal frequency bears on how useful the measure is for a given use. Fast-moving phenomena like flooding or wildfires may require aerial imagery taken by drone (uncrewed aerial vehicle) or imagery captured by satellite systems with daily frequencies, while such slower environmental processes as deforestation or crop growth need only weekly, monthly, or yearly frequencies.

Social data have an extent and resolution as well. Census data are often collected every ten years, while more focused surveys (in terms of the number of people questioned) are conducted more frequently. In the United States, for example, the decennial census is more representative of the population because it captures the characteristics of most of the people in the nation. However, it asks fewer questions than other surveys and has a coarser temporal resolution. Scholars looking at human–environment relationships in health, for example, use these data but can face limitations based on how a given set of neighborhoods may change demographically more quickly than indicated by a snapshot of these regions taken every ten years. In contrast, big data from geotagged social media offer the potential to create more current and even more nuanced snapshots of neighborhoods (Poorthuis 2018). The temporal extent of data describes the period in which they have been collected.

Attribute scale concerns how distinctions are made among categories or classifications. Generally speaking, data with many categories tend to be more detailed and have higher attribute resolution. The age of individuals matters when looking at human-environment issues such as disease risk. If data on susceptibly or impact present only two categories, such as individuals who are under 50 years old and those over, these data have low attribute resolution compared to those splitting ages into more categories, such as data based on five-year intervals (e.g., 0–5 years old, 6–10 years old, 11–15 years old . . .). Attributes can also have extent, although it is a looser concept in some ways than temporal or spatial extent. It can refer to the population sampled, for example, or the range of observations, saying that age data were collected only for people aged 18 years or older. Attribute extent can also relate to the breadth of observations captured over their possible range. The attributes of remotely sensed imagery are determined by the kinds of spectra that they measure, as explored in Section 2.2.4.

Bear in mind that spatial, temporal, and attribute domains interact. For example, since spatial boundaries of regions are often arbitrary, they can change over time in ways that can affect measured or reported attributes. Average snowpack depth can vary for reasons having nothing to do with snowfall. The temporal period over which it is measured or the spatial units over which it is averaged can influence reported measures. We delve further into the challenges of space and time in data in Section 2.3 on bias and accuracy and in Section 3.5 on methodological challenges of handling space and time.

Metadata are a kind of attribute data because they are data about data. Most human-environment researchers use data that are repurposed, derived, or collated. As such, it is crucial to consider metadata. Metadata concern who or what collected the data; how the data were captured, classified, or manipulated; measures of accuracy or validity; and basic features such as data model, resolution, or time. Metadata are vital because they allow assessment of how well the data can apply to a given problem, be combined with other data, or be considered fit for a given use.

Accuracy characterizes how well data correspond to reality. Spatial, attribute, and temporal domains all have accuracy concerns. Spatial accuracy centers on how well data capture real-world locations. Data can have poor spatial accuracy for reasons ranging from poor inputs (e.g., scanned maps or blocked navigation satellite signals) to processing that does not use appropriate algorithms. Temporal accuracy is concerned mainly with the data having the correct date for when data were collected (e.g., the timing of vegetative cover data concerning seasons) or whether details about an object are up-to-date with changes in the world (e.g., delays in supposed real-time data). Finally, attribute accuracy is related to how reported characteristics of an object reflect characteristics of the real-world object. For example, even the best survey data can be inaccurate if those interviewed do not respond, misunderstand questions, or intentionally offer false information. Inaccuracies may also occur when using survey data to estimate values for a population, given the necessarily incomplete nature of sampling. We discuss accuracy in greater depth in Section 2.3, which examines bias, accuracy, and error.

Interoperability describes how well different data sets may be combined for some use. Spatial interoperability asks whether spatial resolution or extent match up. For example, given the need to preserve confidentiality about personal information, human-environment health research often uses patient data aggregated to spatial units such as postal codes. These data are often paired with socioeconomic data reported by census tract or gridded environmental data, so the boundaries of the data (their resolution) are rarely the same. Postal code data usually encompass parts of many census tracts, leading to poor interoperability because the regions do not capture aggregated characteristics for the same group of people.

Similar challenges arise with attribute interoperability, which measures how categories or measures match across data sets. Census data code important socioeconomic factors differently. For example, in the United States census, the number and type of categories that map onto race and ethnicity have changed over time, and so too has a respondent's ability to choose more than one category. The history of the census and other official statistics is rich with examples where the nature of questions asked is contingent on cultural, social, economic, and political processes.

Finally, temporal interoperability relates to how well periods match up. Remotely sensed imagery is essential for developing measures of ecosystem function or land change, for example, and variations of a week or two in when data are collected can matter, such as leaf-on versus leaf-off timing in deciduous forests or greening periods in tropical forests. Data science often relies on combining multiple data sets, but questions of spatial, temporal, and attribute interoperability must be carefully addressed in most analyses. Section 4.1.4 explores the effects of interoperability on modeling and theory development.

There are other dimensions in addition to the spatial, temporal, and attribute domains. Technical interoperability refers to technologies like networks and programming that connect databases. Syntactic interoperability relates to the standards around how structured data are exchanged over technological infrastructures, such as document encoding and web services. Semantic interoperability ensures that meaning remains unambiguous as data are shared across users and contexts via ontologies, metadata, and other approaches (more in Section 3.3.2). Section 6.3 examines infrastructure and delves into the complicated business of building semantic interoperability into big data and computing infrastructure.

Pragmatic interoperability refers to the institutional and organizational aspects of data related to performance, quality, and trust at all steps, from gathering through to dissemination and reuse (Janssen et al. 2014). There are many well-studied human-environment systems, but few offer comprehensive, integrated data sets. The Great Lakes region in North America, for example, is the subject of decades of scientific work that has led to the creation of dozens of detailed environment databases. Nevertheless, these collections remain largely unintegrated in ways that impair the possible study of many human-environment linkages (Bassil et al. 2015). Key challenges include access, harmonization, inconsistent spatiotemporal coverage, and unreliable funding.

Data can vary in how well it is *structured*, from unstructured to semistructured to unstructured (Gandomi & Haider 2015). As the name suggests, *unstructured data* are without an inherent framework or data model designed for analysis. Videos, text from web pages or tweets, and uncoordinated temperature readings are examples of unstructured data that are still amenable to some forms of analysis (e.g., simple searching for values). However, they need processing to be turned into semistructured or structured data. One of the most exciting areas of data science relates to advances in converting unstructured data into structured data. The ability to comb through millions of photos to extract bird species or forest types, for example, is a significant leap forward in some ways for the scientific enterprise. As we discuss later, however, such data pose problems, and more generally, much interest in unstructured-to-structured transformations is being driven by the surveillance state and commercial enterprise.

Structured data is what we usually think of as scientific data stored in a database or spreadsheet. Such data has a consistent structure or data model that facilitates analysis. In contrast, semistructured data are still recognizably text, numbers, or images but are often not standardized in such a way as to be easily stored in a database. They can be hard to query and process. Data within census data tables or remotely sensed images are semistructured in that they are machine-readable and have an internally consistent structure. However, such data are typically not amenable to database operations without processing (Section 3.2 looks at data science tasks).

Data models describe how structured data are formatted, stored, and analyzed. Data modeling is a rich topic in its own right. At their most basic, data models represent data entities and their relationships. They can range from conceptual and abstract flowcharts to logical specifications of operations and data types to physical ones that specify how data are stored on hardware (Brodie 1984). There are a few categories of data models commonly used by human-environment data sets. Note that these are amalgamations that combine logical and conceptual models, and in the data-modeling world, similar terms are used in different ways. For example, *object-oriented* can describe an abstract approach to diagramming entity-relationship models, a programming paradigm, a modeling approach, and a file system.

- *Entity models* are information about individual people, locations, or entities. Tabular data are familiar to most human-environment scholars as they are often used in statistical analysis and stored as flat files or spreadsheets. For example, social scientists commonly use *microdata* (information on individuals and households) while dendrochronologists track sample data in tables. These data are typically stored in simple flat files (as simple as a string of textual alphanumerics) or spreadsheets but can also be stored in relational data tables or more sophisticated formats like entity-relationship data models or object-oriented models. They describe individual things or observations in terms of attendant attributes.

- *Graph models* are distinguished from other data models by how the links or relationships among nodes are as important as the attributes of these nodes. Spatial vector data can be structured as geometric or topological graphs, which allow for sophisticated analyses of spatial relationships. Does an area bound a location or entities, as when determining when groups of animals inhabit a given forest? Does a line join another, as when tracking the flow of sediment transport in a river system? Graph data are also used to store and analyze data on people or objects, especially how they are connected via networks. Social media are designed with this purpose in mind. Graphs connect people with links that can be coded, such as married-to or friend-of, that may be analyzed.
- *Grid models* are tessellations where each element takes an attribute value. Human-environment analyses often rely on raster data, the grid-based format used for remote sensing, and other information collected by sensors with picture-sensing elements or *pixels*. These data are amenable to a wide array of analyses that operate on layers directly. They are often tied to mathematical and computational approaches for matrices.

Again, it is essential to emphasize that entity, graph, and gridded data are just generalized data models. They are essentially shorthand used in this book and elsewhere to describe schemas that are partially conceptual and partially logical. The same data or observations can be stored in one or more models or converted among them. It is beyond the scope of this book to dive deep into this fascinating topic. However, Section 3.3 examines specific ways data are logically and physically modeled in ways that help overcome the challenges of big data (for a more in-depth treatment, see Zomaya & Sakr 2017).

2.2 Sources New and Old

We are seeing exponential growth in data being collected from various sources. While there are many characteristic "v"s of big data (Section 1.2), pride of place is usually accorded to their volume, variety, and velocity. Human-environment data offer these features and many others besides. Researchers and others apply these data to many human-environment challenges, including predicting health epidemics, assessing earthquake risk, monitoring water resources, or developing fire-potential maps (Networking and Information Technology Research and Development [NITRD] 2016). These data are robust because they often have spatial and temporal characteristics that tie them to particular places and times.

Data are collected by sensors mounted on satellites orbiting the Earth, drones in the sky, and ground-based networks of people and machines. Of growing importance are data captured by sensors on the move, especially cell phones, and via social media and other Internet-enabled systems, including an array of specialized devices such as video cameras and personal weather stations. These sources enable a new kind of citizen science and feed fast-growing data sets created by modeling and simulating human-environment systems.

In thinking about data, it helps to distinguish between active versus passive and voluntary versus involuntary data. Passive and voluntary data include information from websites where a person consents to their information being gathered. Locational data from cell phones sold to third parties is usually passive and involuntary (Hilbert 2016). Active and voluntary includes citizen science and other purposeful data collection activities (Section 2.3.4 dives into crowdsourcing). In contrast, active and involuntary data are more limited but arguably include collecting survey data as part of a required activity, such as enrolling in a government program or being monitored as a condition for release from confine-ment. There are gray areas between and across these categorizations. As we explore in Section 5.1.4 on privacy and consent, the notion of consent becomes clouded when a person must use an electronic system to pay a bill or consent to a long and dry legal document that few people are equipped to understand.

Data can also be captured, exhaust, transient, or derived (Kitchin 2014a). *Captured* data are explicitly gathered (harking back to capta, Section 1.2) or observed using various means, including sensors, surveys, or experimentation. Conversely, *exhaust* data are side effects of other processes, such as mouse clicks or the duration a page remains open while a person is using a website. Kitchin notes that exhaust data can be *transient*, or information that never gets stored or used because there is simply too much of it or it does not offer value to warrant the time, effort, or expense. Finally, *derived* data are secondary data products developed from captured or exhaust data, such as statistical extracts or some function, like slope from topography.

2.2.1 Locational Data

Spatial data are at the heart of many forms of human-environment research because many social and environmental processes are spatial. *Geocoding* is assigning spatial coordinates to an object's location on the Earth's surface. There are many different ways to geocode locational data. Most locational data traces back to *land surveying*, the process of developing new spatial information from known coordin-ates. Satellite-based systems have joined these ground-based approaches and a range of secondary approaches working from known locations. High-quality location data act as a Rosetta stone that allows databases to connect any spatial data, which is a linchpin for the data science of human-environment research.

Land surveying was the first systematized way used to determine location. Most kinds of human-environment big data owe the knowledge of their spatial coordinates to foundational work in surveying. A surveyor uses several methods to determine the precise position of objects in space by triangulating (using the geometry of angles and distances) from known locations to unknown ones. Surveying has existed for millennia. It likely originated in the need to define land boundaries, pinpoint locations of key landscape features, navigate on land and sea, and help construct monuments and buildings. Surveying as a field blossomed in the 1800s. Many countries embarked on campaigns to map and classify the land as part of the broader push toward exploration and militarization of settler landscapes. This is a labor-intensive process that involves teams of surveyors measuring ground distances with chains or specially designed telescopes called theodolites that are precisely calibrated to provide angles for triangulation.

Global navigation technology has updated the practice of surveying for human-environment data. Surveying and geodesy have also branched out in measuring the Earth's surface, including a range of satellite-based measurements or even radio signals from distant quasars to develop increasingly accurate estimates of the Earth's shape (Bolstad & Manson 2022). The most popular tools are *global navigation satellite systems* (GNSSs). These include the US Global Positioning System (GPS), Russian Global Navigation Satellite System (GLONASS), Chinese BeiDou Navigation Satellite System (BDS), Japanese Quasi-Zenith Satellite System (QZSS), and European Union Galileo. While GPS is often used as the generic term GNSSs, it refers to the American system; this volume uses GNSSs or global navigation system.

For years, collecting field data entailed careful map reading and wayfinding. Now, it usually involves using a handset that works with one of the GNSSs and the constellations of satellites that emit signals that purpose-built computers use to triangulate position. Global navigation satellite systems originated as military technology and had a period where civilian use required complicated and expensive equipment, but now many devices include global navigation capabilities, including phones, cameras, and vehicles. This availability has driven the spatial capabilities of big data and data science.

Locational data from global navigation systems have drawbacks. Since navigational handsets rely on line-of-site signals from satellites over 20,000 km away, reception can be blocked or degraded in some landscapes, such as urban areas, and work less well indoors. For example, GNSSs offer good accuracy outdoors, down to a few meters, depending on many factors. It can be unreliable indoors or in urban areas where a mix of building heights and materials confuses the reception and interpretation of satellite signals. Global navigation technologies can offset these drawbacks by using base stations or sophisticated handsets that can use more than one network. However, these fixes can be expensive and are not uniformly accessible around the globe.

Cell phone manufacturers developed methods to address the drawbacks of global navigation systems. Mobile devices can sense and triangulate wireless access points and signals to provide locational data. These data may only work well in areas carpeted with known access points, leading to poor locational data for rural areas or areas with limited wireless networks. Near field communication ties a phone to within a meter of a location, such as a supermarket payment terminal. However, these data are usually tied to proprietary networks defined by use, as with payment systems or theft-deterrent systems deployed by firms. There is ongoing research on even more subtle issues, such as where and how a person carries a smartphone on their body can significantly affect or impair measurements (Presset et al. 2018). In sum, there are many potential sources of locational data, but each has its pros and cons. Overall, locational accuracy can vary according to the method used to establish a user's position in space. All methods have tradeoffs that may require these methods to be used in combination (Birenboim & Shoval 2016).

In addition to ground-based approaches, many spatial information systems have algorithmic means of gleaning information. The basic surveying frameworks still serve as ground truth, but remotely sensed imagery and spatial information in *geographic information systems* (GIS, a computer system for spatial data and mapping) and other database systems can almost automatically tie data to their ground coordinates. Satellite data are usually processed by their producers to have the proper spatial coordinate system. Analysts can manually or algorithmically tie a street address or postal code to their ground coordinates in an existing database or interpolate from other coordinates.

Many forms of data have geolocational information embedded in them as metadata. Many of the studies that use social media, for example, benefit from how location is embedded in text and photos. These data are not without problems. A subset of these data typically carries accurate geolocational information. Some services, like Twitter, have changed how they handle locational data in terms of resolution and privacy. The impact of such changes may be ameliorated, however, by the fact that spatial cues may still be extracted from the actual text of tweets or posts, such as activities, specific or generic locations, times of day or season, or other ambient geospatial information (Weidemann et al. 2018). Sections 2.2.5 and 2.2.6 on mobile, social, and transactional data dive into these issues.

2.2.2 Earth Surface Data

The oldest and most consistent data source for human-environment research is the observations made by people and machines on the land and oceans. These data describe a broad range of social systems and features of the atmosphere, hydrosphere, biosphere, lithosphere, and cryosphere. Data capture on the ground is usually done by an individual using their senses (sometimes augmented by technology) or by measuring mechanisms interacting with the world in a given location. Researchers have long collected or used surveys of people in their homes, workplaces, or other settings; Section 2.2.3 covers surveys and administrative data. People also increasingly use cell phones, social media, and crowdsourcing to understand social and environmental systems, as explored in later sections.

Humans have observed Earth systems for centuries. Sensing systems can focus on a single phenomenon, such as rainfall or stream depth, or collect a broad suite of measures designed to capture a wide array of meteorological and atmospheric data. Data collection relates to almost any conceivable system or issue, including agriculture, biodiversity, ecosystem health, climate, weather, energy, disasters, health, land use and land cover, and water. Consider the example of a single mountain landscape, which can be measured via dense sensor deployment that

includes: (1) mobile stations, (2) high-resolution conventional weather stations, (3) full-size snow/weather stations, (4) external weather stations, (5) satellite imagery, (6) weather radar, (7) mobile weather radar, (8) stream observations, (9) citizen-supplied observations, (10) ground [Light Detection and Ranging] LIDAR, (11) aerial LIDAR, (12) nitrogen/methane measures, (13) snow hydrology and avalanche probes, (14) seismic probes, (15) distributed optical fiber temperature sensing, (16) water quality sampling, (17) stream gauging stations, (18) rapid mass movements research, (19) runoff stations, and (20) soil research.

(Lehning et al. 2009, p. 46)

We will examine a broad array of measurements and sensor systems throughout this volume.

Data are collected on the oceans by surface buoys, networks of underwater sensors, and other systems. The Ocean Observatories Initiative (OOI) describes data collected by systems from the surface to the seafloor, measures related to currents, water and sediment plumes, ocean chemistry and biogeochemistry, temperature, and biological features (Smith et al. 2018). These data are used by hundreds of oceanographers and marine biologists in their work and by others, including global climate and ocean modelers. The older Global Ocean Observing System (GOOS) was envisioned in 1991 by the Intergovernmental Oceanographic Commission (IOC) as a way to support a global network of fixed and mobile sensors in space, air, land, and ocean (Moltmann et al. 2019). Deep ocean observation is partly accomplished by remote sensing (Section 2.2.4) and long-running ship-based oceanic surveys complemented with a growing array of mobile sensors deployed on gliders and drones. The latter are increasingly autonomous underwater vehicles (AUVs) or uncrewed surface vehicles (USVs), the water-based analog to aerial drones. Initially focused on climate and weather, uses of ocean observations have expanded into regional and global ocean assessments, ecosystem services assessments, fisheries management, and a range of services related to ocean acidification, tsunami detection, and biodiversity.

Environmental sensor networks are increasingly central to collecting land, ocean, and lower atmosphere data. On-ground sensing for environment data has jumped from automated logging systems that entail manual downloading via a cable or removable drive through to automatic downloading via networks to intelligent networks of sensors and communications systems that can collate and even curate data on the fly (Hart & Martinez 2006). Weather observation, for example, moved from human observers recording on paper what their senses and simple measuring instruments were telling them to automated logging systems in the early to mid–twentieth century to sensing platforms that can share their results in real-time.

Environmental sensing platforms can range from minor to large-scale systems, from a single buoy on a lake or stream gauge to a nationwide hydrological sensing network composed of hundreds of stations. Consider seismological sensing networks. Their origin and development were tied to the Cold War and the desire on the part of many countries to monitor for nuclear explosions related to testing and arms control (Barth 2003). Today they are among the most sophisticated networks globally, with well over a hundred monitoring stations that span from the South Pole to Siberia and down to the depths of the Pacific Ocean. These data are easily accessed via the Internet in real-time and archived via the Incorporated Research Institutions for Seismology (IRIS) data management system (Butler et al. 2004).

Sensor networks assume institutional trappings as surface measurements are increasingly collected as part of large-scale infrastructure. Investments by national funding agencies are typically necessary to create and maintain these networks. Given that many human-environment phenomena transcend borders, there are also international efforts. For example, the Global Earth Observation System of Systems (GEOSS) is a community of nations, institutions, and other stakeholders with the primary goal of coordinating Earth observation platforms to examine globally significant human-environment dynamics (Lautenbacher 2006). It coordinates activities and offers standards but relies on members to integrate their sensors into the collective enterprise (Global Earth Observation System of Systems [GEOSS] 2005). These large-scale projects highlight issues related to human-environment data, including increasing heterogeneity across and within scientific fields due to networking, integration of data from many disciplines, and the impact of new technologies in expanding the breadth and depth of environmental data (Karasti & Baker 2008). Section 6.3 delves more deeply into sensor networks and infrastructure's institutional and computational characteristics.

Data can be captured in new ways with existing sensors and used in unanticipated ways. A fascinating example of developing new data is the use of cellular towers to assess rainfall, where the signals shared among towers are attenuated or degraded by rainfall in ways that give solid estimates of precipitation in areas with adequate tower coverage (Overeem et al. 2016). For another example, seismic data was used to measure the decline in human activities in 2020 due to COVID-19. Pandemic responses such as shutdowns and the attendant reduction in social and economic phenomena led to a global-scale quieting (Lecocq et al. 2020). Perhaps even more interesting is the idea that the seismological measures correspond to other measurements of human mobility. Seismology could provide near real-time estimation of human activity. In some places, readings can be localized to schools or resorts and measured to the nearest minute. In contrast, in others, these measures are broader but still potentially useful for a range of applications beyond those customarily considered.

High-quality ground stations are increasingly joined by lower-cost ones that promise reasonable data quality with good temporal and spatial coverage. Weather enthusiasts have long enjoyed personal weather stations. These devices are increasingly being used in an aggregated way because people can link their stations to weather monitoring networks. These home-based consumer devices commonly measure temperature and barometric pressure and collect data on other atmospheric features, including carbon dioxide and noise levels. There are limits to the data collected by consumer-grade technologies, mainly because scientists do not know critical features of sensor placement, such as height above ground or exposure to airflow, affecting the accuracy of measurements collected (Chapman et al. 2017). Nonetheless, when used in sufficient quantities and when allowing for sometimes poor placement, they can be used to collect simple measures such as temperature and pressure, and are getting better at capturing more sophisticated phenomena.

Another form of in-situ data collection with ramifications for data science is the increasing availability of low-cost air pollution sensors. Many countries and areas employ static air quality monitors to sense carbon monoxide, nitrogen oxides of various kinds, ozone, and particulate matter. These devices offer acceptable attribute resolution and accuracy but can cost tens of thousands of dollars and require attention, calibration, and maintenance. There is interest in low-cost or lower-cost sensors that can be used to blanket a region to provide greater temporal and spatial coverage. One ongoing challenge to using these sensors is that the majority are not accurate enough for scientific research, although they may be adequate for raising awareness about air quality (Castell et al. 2017). More research is necessary to understand how these sensors may be best used and how to accommodate how they are affected by weather patterns, relative humidity, temperature, site, and situation.

Network-connected cameras are an exciting source of human-environment data. Cameras are not new. Biologists have used camera traps and microphones for decades to record a range of animal behavior (Burton et al. 2015). However, what is new is the attendant growth in machine learning and human crowdsourcing to interpret these observations, as explored later. Another fast-growing source of human-environment big data is outdoor internet-connected cameras, or webcams that offer live images or video feeds over the Internet. Cameras have long been used to remotely monitor environmental systems, ranging from coastal landscapes and erosion to phenological changes in vegetation (Román-Rivera & Ellis 2019). Jacobs and others (2009) offer one of the earliest efforts to discover and organize a massive global sensor network. They found that it offers a potentially helpful alternative to satellite imagery and other sensor constellations for global-scale remote sensing (Section 2.4). Environmental data gleaned from webcams include many weather variables (e.g., rain, cloud cover, insolation) alongside deriving information on environmental features such as tree cover, topography, atmospheric haze, and snow depth via automated scene detection. Of course, as with other aspects of data science, cameras have potential drawbacks as a data source, including the proliferation of surveillance (Section 5.1.3) and erosion of privacy (Section 5.1.4).

2.2.3 Administrative Data

Administrative data are a vital source of data for data science. We are seeing the vast discovery, expansion, and codification of existing data, such as census data and other forms of so-called administrative data, which are not always considered big data but are an essential data source for human-environment scholarship. For decades and centuries, much of these data have been locked in physical form in discrete locations, usually books on library shelves. However, administrative data are becoming widely accessible as institutions increasingly seek to digitize, compile, and share these analog resources. As these data sets grow in size and become amenable to data science approaches, they offer a kind of big data that may not always satisfy all the "v" criteria of big data (Section 1.2). However, they are nonetheless an essential component of human-environment research. At the same time, there are tensions within and among scholarly and governmental communities around big data and administrative data.

Many existing administrative data sources are helpful forms of big data. Well-crafted surveys conducted over years make it possible to "truly understand the big picture, whether across large stretches of time or large expanses of territory" (Sobek et al. 2011, p. 61). Long-term surveys offer deeper attribute data in the forms of dozens and often hundreds of variables for carefully chosen questions over long periods. The General Social Survey (GSS), for example, provides a large amount of data on Americans back to 1972. This survey asks a range of questions of the nation's adult population on various behavioral and demographic features.

Survey data are often more helpful than big data because they are relational, aggregated, multilevel, or merged (RAMM). "Relational data have embedded linkages between individuals and groups; aggregated data include data and variables that are combined from multiple sources and measures; multilevel data combine measures at both individual and group levels; and merged data are combined from different original sources" (White & Breckenridge 2014, p. 334). Unlike many big data sets, RAMM data offer a kind of complexity in how they capture the nuance of people's lives and offer a complete and robust view of a range of social phenomena. The same can be said of many traditional forms of environmental data that use smaller-scale studies to advance knowledge of many human and environmental systems (Section 4.3 looks at linking small and big data).

Projects like the Integrated Public Use Microdata Series (IPUMS) offer detailed census microdata on individuals, households, communities, and broader environments. The Integrated Public Use Microdata Series has data on over 1 billion people and looks to have data on 4–6 billion within a decade (Ruggles 2014). These data fulfill many of the definitions for the big data examined in Section 1.2 because they are voluminous and apply to large swathes of the population. They offer variety, complexity, and relationality in how they can capture important structures of the phenomena they measure, such as familial relationships. They are valuable because they often exhibit exhaustivity by measuring the entirety of a census population at fine-scaled spatial and attribute resolution. They are also significant in how thousands of researchers use them across the human and environmental sciences to drive a broad range of domains to drive science, policy, and learning.

Administrative data are often seen as fundamentally different from big data but there are important similarities. On examination, administrative data have many features that make them relevant to the conversations around big data (Kitchin & McArdle 2016). The critical distinction is that administrative data are usually collected during the administration of taxes, loans, businesses, and other contexts in which people interact with the state. Indeed, some nations are steadily doing away with large-scale surveys and relying on administrative data (Struijs et al. 2014).

The leading statistical agency in many countries is the *national statistical office* (NSO). They usually have another name in their countries, but the generic term is used to describe these organizations (Brackstone 1999). National statistical offices have developed principles for large data sets that give insights into the differences and similarities between administrative and big data. Early work on these principles was conducted by NSOs and related bodies, including the International Statistical Institute (ISI), the International Association for Official Statistics (IAOS), and bodies such as the United Nations Statistics Division (UNSD) and the European Statistical System (ESS). The General Assembly of the United Nations adopted a core set of principles in 2014 (United Nations 2014), but their origins can be traced back to the 1990s (Carson 1998). It is challenging to meet all of these principles in practice. However, they are helpful aspirational goals and are a framework for comparing administrative data to other kinds of big data. In simple terms, the relative strengths and weaknesses of big data and administrative data stem in part from their underlying ethos (Table 2.1).

In addition to clear overlaps and synergies in approaches, there are ongoing tensions and attendant conversations around whether big data will supplant or complement more conventional forms of data. The science of (big) data is an object of institutional and disciplinary contention, especially regarding which people or organizations are the sources of authoritative or official data, particularly for the state (Grommé et al. 2018). Researchers in NSOs are concerned about calls to use new data sources such as cell phones or high-resolution remote sensing to complement and ultimately supplant existing administrative data surveys. In particular, there is growing interest in discontinuing censuses and other governmental data collection programs in the face of the potential for big data to create seemingly similar data sets at less expense. National statistical offices have long been the final word in providing quality data for various domains. Moves to circumvent this role or provide inadequate data rightly concern them.

Statisticians working in NSOs face challenges from some kinds of data science. Central to these challenges is the perception that mainstream big data sources, such as cell phone data or social media, potentially offer more data and more timely data (Ruppert 2013). The United Nations Economic Commission for Europe speaks of the pressure to accommodate data science, arguing that it

is unlikely that NSOs will lose the "official statistics" trademark, but they could slowly lose their reputation and relevance unless they get on board. One big advantage that NSOs have is the existence of infrastructures to address the accuracy, consistency, and interpretability of the statistics produced. By incorporating relevant big data sources into their official statistics process, NSOs are best positioned to measure their accuracy, ensure the consistency of the whole systems of official statistics, and provide interpretation while constantly working on relevance and timeliness. The role and importance of official statistics will thus be protected.
(UNECE 2013 in Grommé et al. 2018, p. 58)

Section 2.3.5 explores how big data and administrative data are used in complementary ways.

Table 2.1 *Administrative data versus big data*

Principle	Administrative data	Big data
Relevance, impartiality, and equal access: Official statistics that meet the test of practical utility are compiled and made available on an impartial basis by official statistical agencies to honor citizens' entitlement to public information.	NSOs can offer great relevance and impartiality, although there are instances of political, economic, and social interference with the functioning of official statistical agencies.	Big data are sometimes relevant, are often not impartial, and offer varying access levels. Data science holds out the promise of gathering and creating data that is more relevant for many uses (Prewitt 2010).
Professional standards, scientific principles, and professional ethics: To retain trust in official statistics, the statistical agencies need to decide according to strictly professional considerations, including scientific principles and professional ethics.	NSO officials work within the larger framework of governments and policies. There is pressure on some officials in some places to modify data or obscure procedures.	Big data have some standards stemming from common practices but have been the source of troubling ethical lapses (more on policy in Chapter 5). There is a growing recognition of the need to design systems for privacy.
Accountability and transparency: Statistical agencies are to present information according to scientific standards on the statistics' sources, methods, and procedures.	Most NSOs offer documentation on how they collect and manipulate administrative data.	Some data scientists ardently support open data and methods. Large swaths of big data are still proprietary or are otherwise inaccessible given the significant resource demands and lack of necessary expertise.
Prevention of misuse: The statistical agencies are entitled to comment on erroneous interpretation and misuse of statistics.	This provision does not bear on the vast amount of data collected, used, and misused by intelligence services.	Scholars or firms can comment on how their studies are used, but complaints may have less authority than those from an NSO. Once big data exist, often anyone can use them for nearly any purpose.
Sources of official statistics: Data for statistical purposes may be drawn from all sources, be they statistical surveys or administrative records.	Big data usually draw on a broader range of sources than administrative data.	Big data relies on access to proprietary data versus closely guarded administrative data in the same way that statistical agencies do.
Confidentiality: Whether referring to natural or legal persons, individual data collected by statistical agencies are strictly confidential and used exclusively for statistical purposes.	There are ongoing debates within NSOs and beyond on how to preserve confidentiality best while also ensuring data are usable by policymakers and scholars.	Confidentiality remains a challenge for big data. Many features of big data for business, in particular, rely on identifying individuals and commodifying this information.

Table 2.1 (*cont.*)

Principle	Administrative data	Big data
Legislation: The laws, regulations, and measures under which the statistical systems operate must be made public.	Open legislation is essential but does not preclude lousy law.	Legislation and laws dealing with big data lag practice. As a result, big data often lie mainly outside of governmental influence.
National coordination: Coordination among statistical agencies within countries is essential to achieve consistency and efficiency in the statistical system.	Degrees of coordination can be subject to laws prohibiting data sharing or bureaucratic infighting.	There are some big data analogs in the sense of a shared scientific enterprise, although they compete with commercial or national security imperatives.
Use of international standards: The use of international concepts, classifications, and methods by statistical agencies in each country promotes the consistency and efficiency of statistical systems.	Implementation varies around the globe.	Similar to issues around coordination, there are no standards beyond forms of scientific or commercial practices.
International cooperation: Bilateral and multilateral cooperation in statistics contributes to the improvement of systems of official statistics.	There are many ongoing forms of cooperation, often under the aegis of the United Nations, World Health Organization, or other international organizations.	There are some international movements around using big data for human-environment research.

(Text in the Principle column is mostly verbatim from United Nations 2014).

2.2.4 Eyes in the Sky

A vast and growing reservoir of human-environment data is being collected via *remote sensing*, which involves gathering and analyzing data acquired by a platform at a distance from the Earth's surface. Section 2.4 explores these and other applications of remotely sensed imagery. However, it is helpful to establish some basics here since later sections discuss various aspects of remotely sensed imagery. People have developed many different platforms for capturing images, including affixing cameras to pigeons or balloons and tossing specially equipped sensors into the air.

Today, the two most common forms of remote sensing involve placing cameras and other sensors onto satellites orbiting the globe or on aerial vehicles, including drones and airplanes. These platforms measure electromagnetic spectra ranging from ultraviolet and visible light to infrared and into longer wavelengths corresponding to radar and microwave emissions. These data are then processed and analyzed in ways that make them helpful in addressing human-environment issues and challenges. These include rural development, disaster mitigation and preparedness, environmental security, agricultural productivity, species habitat preservation, resource conservation, carbon cycling, and ecosystem service provision.

Remote sensing has used many of the essential tools of data science for decades. Human analysts have long interpreted imagery with the naked eye and varying mechanical and digital approaches (Jensen 1986). Raw data are typically a measure of electromagnetic radiation. The sensor measures the intensity and nature of reflected sunlight (or laser or radar returns or infrared radiation emissions) captured by each sensing element in the imaging array. A single image can be considered a digital photograph composed of millions of individual pixels, short for pixel elements. The pixel is (usually) a square unit. Its shape is due to most satellites using a regular grid of imaging sensors, and each pixel is the output of a sensor. Translating these disparate flecks of color or shade into something meaningful is easy for a human. We can quickly interpret colors, shapes, and textures and identify features such as water bodies or buildings.

Remote sensing relies on computers for much of its classification and interpretation. Like most forms of data analysis, computers are good at some tasks but less suited to others. Many of the approaches in data science have analogs in processing remotely sensed imagery that take raw pixels and use their spectral, spatial, and temporal attributes to classify them as given kinds of land cover or land use. *Land cover* is the physical phenomena on the ground, while *land use* describes the human uses to which people put land cover. For example, deciduous versus coniferous forests are kinds of land cover. In contrast, land uses are the human activities or values tied to the land, such as logging trees or leaving the forest intact to clean water flowing to a reservoir. These approaches are incredibly sophisticated and predate big data methods by decades, so there are some natural synergies in moving classic methods into a big data context.

Sensors on modern remote sensing platforms capture electromagnetic spectra ranging from ultraviolet and visible light to infrared, along with longer wavelengths, including microwave and radar. Spectral range largely determines what may be discerned with remotely sensed imagery and the kind of attribute data developed from these images. Spectral range refers to the breadth of the electromagnetic spectrum that the sensor captures. A multispectral sensor measures many portions (usually termed bands) of the spectrum (e.g., several different infrared wavelengths). In contrast, a *panchromatic* sensor measures a single electromagnetic band, yielding something akin to a grayscale image. *Hyperspectral* imagers are multispectral sensors that capture dozens or hundreds of bands, each spanning a narrow slice of spectrum. These data can be used to delineate subtle variations in leaf water and nitrogen content in vegetation, differentiating among vegetation types, such as picking out invasive species among native ones, or other measures such as biomass and plant health (DeFries 2008). Other sensors are designed to capture fine temperature gradations such as those due to the urban heat island effect or different types of geological strata (Lo et al. 1997).

The majority of remote sensing platforms inhabit the visible and near-infrared parts of the electromagnetic spectrum, which map more or less onto how humans perceive color in the landscape. These data can be used to characterize the Earth's surface by combining spectral bands. For example, the normalized difference vegetation index (NDVI) blends several bands to measure vegetative activity and state, including biomass, leaf area, and tree type. Infrared bands capture the thermal range of the spectrum, namely heat of the Earth's surface, which may be used to measure moisture availability, evapotranspiration, and variations in surface materials, such as concrete versus soil. Visible and near-infrared data have been collected for the entire globe since the 1970s at spatial resolutions of dozens of meters down to below a single meter more recently. These data are voluminous and well understood, although data scientists are opening up new vistas by drawing on vast amounts of the data (Faghmous & Kumar 2014b).

Nighttime imagery was once primarily the domain of military intelligence (e.g., seeking troop movements or fuel depots), but such data are now firmly in the civilian sphere. Artificial nighttime illumination is valuable for capturing human activity, such as demographic movements, and offering insight into subtle features like linking electrification to poverty measures. For example, researchers used Defense Satellite Meteorological Satellite imagery to track changes in patterns of nighttime lights in Baghdad, Iraq, over the years 2003–2007. They assessed the impact of war on population, electrification, safety, and sectarian violence (Agnew et al. 2008). Other examples include tracking energy usage relative to broader economic changes (Filho et al. 2004) and assessing the human footprint in natural areas (Levin et al. 2015).

Sensing platforms can use microwave and laser emissions. The microwave portion of the electromagnetic spectrum may be measured by both passive systems (those that read energy that is naturally emitted or reflected from the landscape) and active systems (radar) that bathe the landscape with their energy and read the reflections (Thenkabail 2019). Active microwave sensors can penetrate cloud cover, making them ideal for tropical or coastal regions. They also measure physical surface characteristics (e.g., shape, roughness, and orientation of features) and surface reflectivity and conductivity, useful for atmospheric water content, soil moisture, land cover, and surface height and topography. One of the earliest and still valuable data sets is the Shuttle Radar Topography Mission (SRTM), which used radar to develop a global-scale map of elevation moderate (30 m) resolution (Rabus et al. 2003).

Light Detection and Ranging (LIDAR, more commonly lidar) is another crucial source of big data for human-environment research. Lidar involves beaming pulses of laser light from a transmitter and collecting returns from the surface. It is primarily used to generate high-resolution three-dimensional surfaces, essential to various applications, including assessing solar electricity potential, flood monitoring, and silviculture inventory. Lidar points are typically collected at scales of 1 m and have a vertical accuracy of $\pm$ 0.10 m, which allows the creation of an elevation model with submeter resolution and accuracy (Wang & Allen 2008).

Spatial resolution has much to do with how imagery poses big data challenges (Bolstad & Manson 2022).

- *Moderate-resolution data* form one of the largest satellite imagery collections, offering 10–250 m resolution (these measures typically refer to the length of a pixel's side). The earliest platforms date to the 1970s and include Landsat Multispectral Scanner (MSS) (with a spatial resolution of 79 m), Landsat Thematic Mapper (TM; 28.5 m), SPOT (Système Pour l'Observation de la Terre) imaging instrument (10 or 20 m), European Sentinel (10–80 m), and Indian Remote Sensing satellite (IRS, 18–30 m, although newer iterations include fine-scaled imagery as well). Moderate-resolution imagery captures land cover change, habitats, and changes in land use.
- *Coarse-resolution imagery* trades time for space, capturing spatial resolutions of 250–1,000 m or more in exchange for temporal frequencies of a day or so. Platforms offering coarse-resolution imagery include the Advanced Very High Resolution Radiometer (AVHRR; 1.1 km resolution) and Moderate Resolution Imaging Spectroradiometer (MODIS; 250 m or 500 m resolution).
- *Fine-resolution imagery* is captured by a growing array of platforms with a spatial resolution of 0.1–10 m. These include Satellogic ÑuSat (1 m), QuickBird (0.6 m), GeoEye (0.4 m), and WorldView-2 (0.4–1.8 m). Fine-scaled platforms increasingly offer temporal resolution on the order of days because they use collections of multiple satellites. While some of these products, like the newest SPOT and Gaofen (China) satellites, are publicly owned, there has been explosive growth in private firms launching constellations of microsatellites housing relatively low-cost sensors. Each satellite and sensor is not particularly powerful or as advanced as the instruments on larger vehicles. However, they offer good spatial resolution, excellent spatial extent, and high temporal frequency in aggregate.

In addition to improvements in the basic sensors themselves, or the introduction of new sensors, technological integration is advancing data collection. While imaging systems are becoming increasingly powerful because of technological advances, they also gain from converging improvements in different technologies.

Global navigation satellite systems and inertial measurement units (devices equipped with accelerometers, gyroscopes, and magnetometers) offer precise, accurate, and virtually instantaneous information on the position (attitude of the platform) and location of sensing platforms (Section 2.2.1). These measures give the resultant imagery great location accuracy and aid in post-collection processing. These data may be enhanced with lidar-based elevation data. While many of these data sources are brought together after collection, they are increasingly combined during collection, saving much effort in analysis (Campbell & Salomonson 2010).

2.2.5 Data on the Move

Mobile devices offer new ways to use data science to explore human-environment systems. The widespread adoption of global navigation and related locational technologies (examined in Section 2.2.1) has spurred a revolution in collecting and using spatially aware data. At the same time, there is a growing array of technologies that allow for the collection of attribute data, including sensors on cell phones, cars, and agricultural equipment. *People-centric sensing* is now a vibrant research and commercialization area, combining "several research disciplines, including sensor networking, pervasive computing, mobile computing, machine learning, human-computer interfacing, and social networking" (Campbell et al. 2008, p. 14). Tablets, computers, watches, and especially phones collect various data, including pictures, videos, audio snippets, time, location, speed, direction, and acceleration. The common use of mobile devices in research dates back to the late 1990s.

Mobile devices offer high-resolution spatial and temporal observations. Compared to other forms of monitoring people through space and time, such as travel diaries, these records are more accurate and do not require research subjects to rely on memory. While devices and contexts will vary, their data offer spatial resolutions of meters to a submeter and data with near-continuous measurements that can be made over long periods. Hundreds of studies have been conducted on the use and utility of mobile devices to understand human mobility. Data are derived from location and a range of sensors in research domains spanning Earth sciences, geography, health, tourism, and transportation (Shoval et al. 2014). There is a good deal of work on using locational data from different modes of travel to better capture commuting and traffic flows that translate into measures of migration and environmental impacts of air or noise pollution (Hong et al. 2020).

The move from cell phones to smartphones in the mid-2000s opened up even more expansive data vistas, as respondents could enter information in applications

Table 2.2 *Cell phones and sensor sources*

Measure	Sources
Location	GNSSs, cell tower location, cell tower triangulation, wireless triangulation from known base stations
Proximity	Bluetooth and radio frequency identification (RFID) can tie the cell phone to other phones and devices with known locations, such as doorways that can read RFID tags or payment locations with Bluetooth
Motion	Determine velocity and types of motion (e.g., walking versus driving), accelerometer for velocity, gyroscope for orientation in space
Electromagnetic	Magnetometer for direct, light sensors, microphone/acoustic, camera
Environmental	Barometer to sense pressure (a proxy for elevation and weather), thermometer, humidity sensor

or *apps*. The phones could provide various measurements (Birenboim & Shoval 2016). Per Table 2.2, cell phones offer a surprisingly large number of measures and sensors for an everyday consumer device. Apps can also collect real-time responses from people. An early example is an app that occasionally asked phone users how they were feeling as a measure of subjective well-being and linked their responses to locational information to determine that people generally felt a greater sense of well-being in natural environments (MacKerron & Mourato 2013). Not all phones offer all sensors, but as manufacturers adopt increasingly advanced chipsets, these sensors will become more prevalent. Finally, scholars use geotagged photographs (cell phone photos are often encoded with geographic location) to derive valuable human-environment data. As noted in Section 2.2.2 in the case of webcams for Earth systems data, photos can indicate surficial and hydrological characteristics such as water clarity and land cover type. Photos can be stitched together and assessed to derive physical geography features (Smith et al. 2016) or people's more qualitative experiences of the natural world (Jones et al. 2011).

We can distinguish between participatory and opportunistic sensing. In the former, as the name implies, users actively engage with their phones via apps, such as taking photos, reporting their feelings, or answering survey questions (Lane et al. 2010). Opportunistic sensing relies less on users and more on phone processes and sensors that run in the background. There is no agreement on whether participatory or opportunistic sensing is generally better for research. Like most methods, it depends on the research questions being asked. Opportunistic sensing offers the advantage of not requiring user interaction or intentionality. In contrast, participatory sensing can offer outstanding detail and specificity when users fill out surveys or answer questions that can then be tied to location and other environmental contexts.

Both participatory and opportunistic data may be tied to the concept of sousveillance, a reversal of surveillance, where the French word for over, or *sur*, becomes under, or *sous* (Mann et al. 2003), to mean inverse surveillance, where people measure aspects of their existence via technology. Initially positioned as a way to surveil the surveillers, sousveillance is usually seen as monitoring aspects of self, such as health or fitness, through heart rate monitors and motion detectors that track the number of steps taken. These technologies have the potential to collect human-environment data on their own or when combined with mobile technologies.

Like many kinds of data, those collected via mobile devices can be combined with other forms of data. Health studies increasingly use cell phones to track users as they move through the environment. For example, researchers have combined cell phone location records with satellite data and other information to estimate human poverty (Steele et al. 2017). De Nazelle and others (2013) offer an interesting case study on using cell phones to complement traditional approaches for assessing human exposure to environmental pollutants. They use a smartphone application to track a person's time, location, and physical activity patterns and compare them against pollution maps. Phones offer a high spatial and temporal resolution that improves exposure estimates. This approach allowed a more nuanced sense of how the activity relates to the environment. For example, the researchers learned that walking accounted for 6 percent of people's time but almost a quarter of inhaled nitrogen dioxide. Keep in mind that this approach only gives better spatiotemporal information on the movement of people. Mobile devices equipped with dedicated pollution sensors would be even better at assessing exposure to pollution.

Precision agriculture is another example of the power of combining data on the move from multiple sources. This form of production focuses on using high spatial resolution monitoring of environmental and crop conditions to guide the application of inputs. It originated in the 1960s with the development of high-quality soil and crop mapping. However, it came into its own with high-resolution data and global navigation systems tied to computerization in agricultural machinery, collecting various measurements with field-based and farm equipment sensors (Sward et al. 2022).

Data science is advancing the ability of agriculturalists and firms to collect and harness a range of data for analysis. Data are collected down to the level of individual plants and plots centimeters wide and are aggregated across farms and regions. The specificity of these data harnessed to near real-time analysis promises to make farming more efficient and profitable. It encourages minimum inputs like fertilizer, and tracks livestock well-being as a function of location, feed, and activity, and moisture content in grain as it is collected to predict better the time needed to dry it for the market (Carolan 2017). Precision agriculture has policy-related advantages and disadvantages, as explored in Section 5.4.

Finally, a corollary of artificial intelligence growing commonplace is that it induces its own data demands and creates its own challenges. A prime example is how automobiles increasingly come equipped with radar, lidar, and optical cameras. They collect data for onboard computers that help people drive under normal conditions, including augmenting cruise control and lane following. They also help in potentially dangerous situations, such as emergency braking in the event of an accident. These sensors capture the movement of other automobiles and people traveling via other modes, such as walking or biking. While these data are usually used in autonomous driving (e.g., to avoid hitting pedestrians and bikers), there is great potential to share these data for scholarly purposes. As manufacturers embrace common standards, connected vehicles should be able to share data amongst themselves. Companies quickly embraced connectivity for autonomous driving to enable platooning in commercial fleets, where trucks can follow each other closely to cut down wind resistance and ultimately reduce labor costs by reducing the number of truck drivers (Xie et al. 2019). These data are not without challenges. Particularly important is that more data exist for automobiles than other forms of transport. This inequality can introduce bias against nonmotorized forms of travel in terms of accurately assessing their use and occupancy (Griffin et al. 2020).

2.2.6 Social Media and Trawling the Net

Social media and transactional data are closely tied to mobile devices that can shed light on human-environment systems. Social media facilitates people interacting over the Internet, while transactional data are the traces of information and the metadata people create throughout their lives when interacting with technology. Researchers use data from a broad array of social media systems, including general ones like Facebook and Twitter, to more narrowly tailored focused ones, like photo-sharing sites Flickr or Instagram. Many services connect people through their location, like Four Square, Nextdoor, or Neighbourly India. Scholars also use transactional data, usually collected by private firms and public agencies, for insight into various social and environmental phenomena.

Social media offer data on coupled human-environment systems. These data can be inexpensive compared to many traditional sources such as remote sensing, even when including the cost of automating analyses, and more widespread than other kinds, such as hydrological or climate monitoring stations. Compared to many surveys, these media do not introduce interviewer bias (e.g., a respondent is not being guided by the survey taker) or suffer from a poor recollection of past events

(Shook & Turner 2016). Many studies focus on social dimensions of human-environment systems, such as using sentiment analysis to assess spiritual or leisure activities in nature (Oteros-Rozas et al. 2018). Others look at more complex issues such as sustainable consumption or environmental impacts of infrastructure development, land use and land cover changes, or climate dynamics such as precipitation and heat waves via textual analysis (Ghermandi 2018)

Social media can offer a rich trove of visual observations on various human-environment and environmental phenomena. Photographs shared via social media are used to give insight into atmospheric and hydrological domains, including weather, precipitation, air pollution, and surface water; along with terrestrial domains, including natural hazard management (Zheng et al. 2018). Researchers can validate radar-sensed hail storms with ground-based photographs (Hyvärinen & Saltikoff 2010). These collections can be analyzed with artificial intelligence methods to extract social and cultural aspects of the environment. They capture data that can assess ecosystem services like aesthetic value access to clean water or ecological indicators like vegetation status, cloud cover, and insolation (Keeler et al. 2015; Lee et al. 2019). Photographs and videos have applications in hydrology, such as when using video to assess water velocity in streams (Le Boursicaud et al. 2016).

Data derived from social media can provide timely observations of events compared to many other sources. Twitter feeds and Facebook posts offer textual information on flora and fauna sightings, earthquakes, hazardous weather, and disease spread (Bordogna et al. 2016). These data can capture fast-moving phenomena like floods or extreme weather that can escape remote sensing or other data collection platforms that use periodic sampling (Fohringer et al. 2015). Data science methods can draw semantic information from tweets to detect thousands of floods over four years for the globe. The resolution is coarse, on the level of a country or first-order administrative regions like states or provinces, but arguably helpful in broadening the information base. This example uses several approaches to extract semantic information from tweets, accommodating how the word "flood" can take on different meanings in different contexts and assessing bursts of activity that are more likely to capture a significant hazard event (de Bruijn et al. 2019).

Social media data have spatial and temporal limits. Simple measures derived from these data, such as the spatial density or temporal intensity of social media posts around the time of a disaster, help pinpoint the location or timing of an event. However, these measures can miss essential attributes, such as the sentiment of people posting or the motivation of people moving through space. These data exhibit spatial bias. Mobile devices are used more often in areas with higher population density or where people can access, such as areas with roads (Stafford et al. 2010). People are more likely to take pictures of charismatic fauna or beautiful scenery, which introduces bias into any assessment of the absence or presence of a given species or the nature of a landscape (Wood et al. 2013). Section 2.3 delves into how differences over space, time, and scale can play out in ways that affect the quality of social media data. Finally, many data sources can disappear overnight, like MySpace or Google Plus, or change dramatically over their lifetimes, such as products offered by the locational intelligence company Foursquare (Noulas et al. 2011).

Social media can have complex temporal dynamics that researchers must accommodate with care. Many events have a seasonality that can bias observations on social media, such as when people spend less time out of doors when it is relatively hot or cold (Arribas-Bel et al. 2015). Day-to-day dynamics or other temporally sporadic events can be obscured when they happen against a backdrop of decreasing or increasing numbers of users. Information about events happening in one locale is quickly picked up and retweeted or reposted elsewhere, making the observed phenomenon appear broader than in reality (Jongman et al. 2015).

There are also feedback effects, such as the rush to take photographs of locales and post them on social media once they become popular in, well, social media. The super bloom of flowers in California, United States, in 2019 is one example. People flocked to take pictures of unusually lush fields of flowers, creating a positive feedback loop where more social media influencers and nature lovers showed up and threatened the very phenomenon that they sought to experience. As one report described, "smartphone-equipped fans, who have shown up in droves over the past three weeks, bringing with them horrible traffic and occasionally horrible etiquette when they wander off the trail to pose with, trample or even pick the poppies" (Stone 2019, p. 2).

Transactional data are a significant source of big data integrated into human-environment research. These data refer to records of transactions or interactions people have with other people, firms, or the state for a mix of big and small events. For example, for centuries, the state has recorded life events from birth until death alongside a range of information related to issues of interest to governance, such as taxation or security, and attendant information on employment, income, and migration. Many organizations track these behaviors, and their scope has expanded to capture a massive array of interactions, events, and transactions (Table 2.3). Internet providers, web companies, and others track every user's click or mouse movement. These are collected whenever a person interacts with any information system, including transit systems, credit card processing, shopper loyalty cards in stores, and banking systems.

Table 2.3 *Transactional data categories and examples*

Category	Types and examples
Life events	Birth, marriage, divorce, and death certificates are readily available in many jurisdictions
Travel	Toll transponders or other radio frequency tags; license plate readers; travel tickets (e.g., train or airline); car rentals; global navigation system receivers in cars, either explicitly (e.g., insurance companies will give you more accurate rates based on actual driving) or implicitly (e.g., cars generate data sent to companies)
Civic life and the state	Voting attributes, such as a person's party affiliation or whether they voted in past elections; real estate and tax information; educational transcripts; driver's licenses; utility bills; tax returns
Commerce	Commercial information, including purchasing and credit information
Health and environment	Apps collect personal information such as heart rate or temperature as part of ongoing care and institutionally collected information such as hospital admissions or wellness tracking. Detailed medical records are shared among private and public agencies, including insurance companies, information clearinghouses, and providers
Web information	Intersects with other forms, especially commercial sources (e.g., online click rates or purchases) and images, text, and videos posted via many applications; social web (e.g., commenting or shared interests such as outdoor activities); gaming and virtual worlds

Government agencies, political parties, advertisers, and internet companies are just a few of many entities that capture, maintain, and distribute vast amounts of information on individuals from before they are born until well after they die. The irony is that often these data are captured and collated with consent. People agree that the government collecting data is necessary to help with services, and they consent to be tracked by firms to get better deals on services, faster route suggestions, and more germane search results. As discussed in Chapter 5, it is increasingly apparent to many people that these data are subject to misuse. Firms underinvest in security and suffer privacy breaches, criminals steal identities to commit crimes, and governments unnecessarily monitor and target certain citizens for potential or actual behavior.

2.2.7 Crowdsourcing and Citizen Science

Crowdsourcing is a fast-growing source of human-environment big data. The definition and nature of crowdsourcing evolve as more people use a growing variety of technologies. As the word suggests, crowdsourcing, at its simplest, means having people contribute data, often via the Internet and smart devices

(Muller et al. 2015). Crowdsourcing is not a one-size-fits-all endeavor. Among the most common forms of crowdsourcing is the use of social media via opportunistic sensing to collect human-environment data, but there are others. As explored in Section 2.2.6, these data can provide information on people, nature, planning, and governance.

Crowdsourcing can be seen as various combinations of people, sensors, and the Internet. In some cases, it involves citizens who use sensors to report data via the Internet (Figure 2.1). In other cases, it involves just people and the Internet, or just people and devices, and so on. Of course, sensors and sensing networks rely on people at some point for emplacement and maintenance. This framework of interacting citizens, sensors, and the Internet drives home that crowdsourcing is a broad category of data gathering that becomes citizen science when people purposely contribute effort to scientific projects.

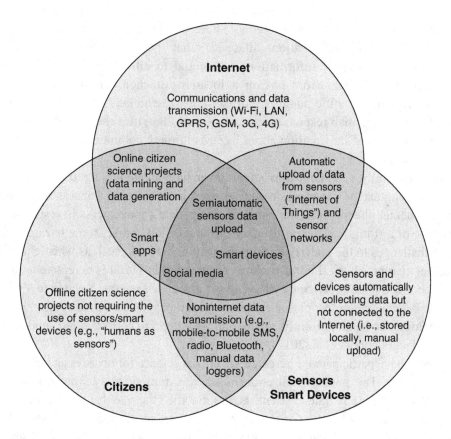

Figure 2.1 Crowdsourcing methods defined by intersections of the Internet, citizens, and sensors/smart devices (Muller et al. 2015). Reprinted with permission from John Wiley & Sons.

There is much potential for integrating crowdsourcing with other methods, including machine learning, remote sensing, and environmental sensors. Such integration can help reduce many socioeconomic biases in social media data by including a broader array of people and groups (Ghermandi & Sinclair 2019). Machine learning and crowdsourcing can be combined to analyze unstructured data from many sources. Examples include using weather radar or bird calls to track migratory birds and their movements, species distribution, and shifts in breeding range (La Sorte et al. 2018). Another example is when researchers bring together experts, members of the public, and computing to create new knowledge about diseases. Much data on the H1N1 influenza pandemic crisis in 2009 was collected via the combination of the Internet and open data, which offered a new way to capture how people were experiencing the pandemic, ranging from access to antiviral drugs and hospitals through to the personal stories of health care providers (Boulos et al. 2011).

Volunteered Geographic Information (VGI) is central to crowdsourcing on coupled human-environment systems. These are locational data (or attribute data with spatial information attached) that people gather and share. Volunteered geographic information is essential to citizen science since data collected often benefit from having a location attached. These data support a wide array of scientific pursuits. For example, researchers can use cell phone data to observe human responses to hazards, including movement through space and propagation of information about hazards through communication networks (Bagrow et al. 2011).

More generally, cell phone movements, even without attempting to ascertain the semantic context of messages, could be used in delineating hazards and other environmental phenomena. The patterning of people's movements in response to flooding or a storm is different from an earthquake. Of course, some hazards will pose challenges to the underlying communication network, such as when extreme weather disrupts physical infrastructure or impairs calls. Hazards in remote regions with poor cell phone service may have fewer associated reports. Related to VGI is *neogeography*, where people untrained in geography or cartography use spatial technologies and expressions like maps and web pages to collect, curate, and share spatial data (Connors et al. 2012).

People have participated in research via citizen science for decades and arguably for centuries. The professionalization of science, turning away from self-taught naturalists to formally trained scientists, has modified the role that citizen scientists play, but this role still exists. Citizen science has a long and distinguished history. For example, in the United States, the National Weather Service's Cooperative Observer Program (COOP) has over 11,000 weather observers who devote more than 1 million hours each year to collecting daily hydrological and meteorological data (Retchless 2018). These observers use high-quality, fixed monitoring stations and share the data via various means, including mail or telephone calls. The public also advances science by contributing to historical data sets, museum collections, and more recent data collection efforts (Cooper et al. 2012). Poisson and others (2019) looked at the contribution of citizen science to lake water-quality measurements in the United States. They found that for decades, citizen science programs produced the majority of observations for essential quality measures, including clarity, nutrient levels, and biomass. This long history demonstrates how cutting-edge data science technologies help deliver valuable data, but an array of powerful alternatives joins them.

Data science tools and technology have increased the role and importance of citizen scientists in human-environment research. As information becomes more widely known and disseminated, scholars increasingly rely on citizen scientist data to better understand long-term environmental change. Specialized apps and websites are at the heart of citizen science, where individuals act as sensors to collect information on weather, crops, animal movement, and plants (Goodchild 2007). For example, flora and fauna identification apps offer opportunities to assess species distribution and abundance, including habitat associations, impacts of environmental change, and movement. These data also give insight into spatiotemporal trends in history and behavior, such as seasonal or decadal shifts in vegetation in response to climate change or shifts in demographic traits such as breeding or animal appearance due to habitat destruction. They also offer rare observations, including infrequent sightings of plants or animals or recording uncommon events such as predation (Cooper et al. 2012).

Among the most successful citizen science networks are those that engage bird watchers. Systems including eBird and iNaturalist are web-enabled citizen science platforms where amateur bird watchers use a web-based system to collect, store, and manage observations. This database helps watchers, researchers, and conservationists link bird patterns to human and environmental correlates (Sullivan et al. 2009). These data were instrumental in assessing the impact of the 2010 Deepwater Horizon oil well blowout and subsequent oil spill because they were immediately available when other research observations were not (Hampton et al. 2013). iNaturalist is a joint initiative of the California Academy of Sciences and the National Geographic Society. It has over half a million members who have recorded millions of observations for more than 100,000 species (Nugent 2018). It uses artificial intelligence approaches to help interpret sightings for error-checking and streamlining observations because crowdsourced data can have gaps and biases (more on data quality and crowdsourcing in Section 2.3.4).

2.2.8 Modeled and Simulated Data

Modeling is a longstanding data source, and its importance is growing for human-environment scholarship. Modeling and simulation of real-world processes are vitally important to many scholarly domains and are central to human-environment research with big data. A full exploration of these data is beyond the scope of this book, so here the focus is on climate modeling to highlight the role of modeling as an information source. Most other fields under the mantle of human-environment research rely to varying degrees on modeling. They will likely see more as part of the broader push toward data science, big data, and computing as ways to describe and understand complex systems.

Climate modeling is a vital tool for understanding many human-environment systems. Simulations are essential to developing short-term weather forecasts, planning for natural hazards like flooding or storms, and assessing the long-term effects of climate change on humans and the Earth. Researchers have built climate models for many years, but these efforts have increased in complexity and sophistication in the last decade. Accompanying this growth in modeling have been significant increases in data, to the point where modeling is on track to be one of the largest sources of human-environment data, from relatively modest beginnings in the 1990s to hundreds of petabytes per year by 2030 (Overpeck et al. 2011, p. 701). As data increases, investments in simulations can integrate recent observational data as data and methods improve. There is exciting work on high-performance and cloud computing that draws on external data sources in simulations and coupling data and models across computing centers (Gogolenko et al. 2020).

Reanalysis involves using modeling to extend or interpolate existing data sets. Climate modeling can integrate existing real-world observations into analyses of previous periods. This method facilitates benchmarking competing models or simulation runs to improve these models incrementally, impute missing values, and create new variables. This work has many existing applications, including mapping hydrological gauge data onto models of streamflow events, vegetation changes, sea ice, and various atmospheric phenomena (Do et al. 2020). These methods have reached a degree of maturity that enables them to be applied to many Earth systems, with the result that reanalysis is creating huge data volumes. Modern-Era Retrospective Analysis for Research and Applications, version 2 (MERRA-2) is a prime example. This US National Aeronautics and Space Administration (NASA) project updates a gold-standard data set and relies on a reanalysis of the modern satellite era produced by NASA's Global Modeling and Assimilation Office (Gelaro et al. 2017). This reanalysis is subtle and important because it helps reduce errors in the underlying observing systems and improve the representation of systems like aerosols, cryospheric processes, and water cycles.

Just as reanalysis seeks to combine and compare multiple data sets and models, so do efforts that compare climate models. Projects like the Coupled Model Intercomparison Project (CMIP) and Atmospheric Model Intercomparison Project (AMIP) seek to compare models and their results to make better models (Touzé-Peiffer et al. 2020). A side effect is the creation of systems to archive and trade the vast amounts of data created by both the models and the comparisons. There has also been extraordinary growth in models at multiple scales, increasingly fine resolutions of meters and kilometers, temporal resolutions of days, and temporal extents out to centuries (Giorgi 2019). To these better spatial and temporal characteristics are added detail and variety in modeled systems, including better cloud cover models and more realistic human-system dynamics. Finally, there is ongoing interest in creating better input data sets for the various modeling efforts, exemplified by the Observations for Model Intercomparison Project (Obs4MIPs), which creates and curates data sets for CMIP on the order of petabytes (Waliser et al. 2020). These model comparisons and attendant data creation efforts join reanalysis and climate modeling as vital data sources and exemplify the growing role of modeling in data science and human-environment inquiry.

2.3 In Data We Trust? Bias, Accuracy, Error, and Uncertainty

Big data from sources old and new are beneficial for human-environment research. However, they can only reach their full potential when scholars consider issues of bias, accuracy, error, and uncertainty. Fortunately, there is a strong foundation of scholarship on the trustworthiness of data that includes considerations of error and uncertainty. More concretely, many questions in data science are about bias in who and what counts. Spatiotemporal gaps are also essential to consider in most kinds of human-environment research. These gaps and other kinds of error play out in crowdsourcing and citizen science, which is steadily becoming a crucial part of data science as applied to human-environment systems. Combining data sources is also a helpful way to combat error and inaccuracy, as seen by combining administrative data with other forms of big data.

2.3.1 Error, Uncertainty, and Trustworthiness

Trustworthiness offers a valuable framework for error and uncertainty in data. Guba (1981) couches trustworthiness in terms of "naturalistic" inquiry-centered qualitative research and mixed methods and "scientific" work in the sense of the normal science approaches to which many data science practitioners would relate.

In this framework, the *truth value* is the fundamental expression of how well data map onto observed or experienced reality. The notion of falsifiability is often used in many scientific disciplines, including data science, to assess the truth. In contrast, naturalistic approaches used in qualitative research seek to establish credibility with people being interviewed or who work on a given problem. This form of truth applies to data science when dealing with methods with opaque inner workings. As explored in Section 4.1, the epistemology of data science is still unsettled, and there are unanswered questions about how to assess the trustworthiness of data and models.

Truth value is tied to other features of error and uncertainty. *Applicability* extends truth value to the ability to generalize data beyond a specific time and place or to which one person's experience holds for another. *Consistency* deals primarily with reliability (of instrumentation, for example, or observations), but it is a precursor to replicability. For more qualitative work, consistency is less about an unchanging observer, such as assuming that an interviewer never changes over time, and more about tracing how relationships with ideas and people affect observations. For example, consistency can increase as the interviewer comes to know the interviewees or subject better. Consistency is also contingent on *neutrality*. Neutrality is often considered in terms of objectivity in scientific research. However, it can also be dealt with by seeking to build confirmability by having multiple observers working with the same material or observations to arrive at some shared result. Bear in mind that qualitative and quantitative are loose labels. As Section 4.2 explores, much nonquantitative work goes into nominally quantitative data science.

Error and uncertainty analyses have a long and complicated history across the sciences, and big data offers new challenges and opportunities. Integrating data sets, often without human supervision, can lead to errors that propagate unexpectedly. Ecological and social data are usually collected according to a careful design meant to support a specific kind of analysis, including conducting power analyses, evaluating sampling regimes, and other tasks to ensure that data will give good results. Combining these data sets requires careful work. Macroecology is home to such efforts, particularly by aggregating many smaller data sets into larger ones. Scholars use approaches like comodeling species (using comprehensive data for one species as a proxy to less-measured species) or using ensemble approaches (using several different modeling approaches to offset gaps in data and or methods) (Wüest et al. 2020). The challenge for data science is that it often uses data for reasons having nothing to do with why the data were collected. This ability to vacuum up and use disparate data sets is valuable for research. However, data science is like any other method in facing limits to how it can accommodate bias, error, and uncertainty.

2.3.2 Who Counts, and Who Is Counted?

A commonly touted advantage of big data is that they offer expansive views on entire populations, but this is often not the case. Scientists spend much time finding new ways to extrapolate from samples to larger populations. So, they are justifiably excited at the potential for big data to reduce or eliminate the need for samples. Alas, it turns out that big data are not a panacea because they are often not representative of many populations. To paraphrase Catherine D'Ignazio and Lauren Klein (2020), who and what gets counted counts in data science. As explored through this volume, samples are biased and incomplete for many reasons. Some relate to race, ethnicity, gender, and class, while reflecting other gaps, including the urban–rural divide, regional disparities within countries, and other socioeconomic differences within and among the world's nations (more on digital divides in Section 5.2).

Cell phones and other devices are among the most commonly used big data sources, but they suffer from biases. Phone use may vary considerably by socioeconomic characteristics, including race, ethnicity, gender, and class. In Rwanda, for example, phone users are disproportionately male, older, and better educated. The amount of time people spend on the phone varies by gender and class. Phones may be shared among people, which confuses the sample frame by making it hard to ascertain who is using the phone (Blumenstock & Eagle 2012). Phone access and use in rural areas reduce the sampling frame such that basic population estimates derived from phones can be less accurate than those derived from remotely sensed imagery (Deville et al. 2014). However, they promise higher temporal resolution data as cellular technology is adopted in rural areas. Locational accuracy can vary according to where photographs are taken, especially in areas with stark relief, such as urban areas or mountainous regions, which affect global navigation systems. There are also gaps between what low-end phones and cameras with quality global navigation system units can produce (Oteros-Rozas et al. 2018). While the share of the world population with cell phones will likely continue to rise, caution is warranted because phone use differs across people and places in meaningful ways.

Social media share many issues that affect the quality of cell phone data. Most commonly used big data sources draw from biased population samples defined by class, space, and time (Li et al. 2013). For example, Instagram, Flickr, and Twitter are three different but commonly used sources of photograph data. However, each creates different world views as measured by spatial and descriptive statistics due to differences among the users of each service (Rzeszewski 2018). These characteristics can also vary over space, from the scale of local regions to countries or entire parts of the globe as a function of access to basic infrastructure, including roadways, electricity, and the Internet. Social media users on mobile devices will have various kinds of biases tied to race, class, gender, and other characteristics (Franklin et al. 2022).

Researchers often have little control over how commercial data are selected, changed, or shared. Social media companies regularly remove material for various reasons, including language, content, or competitive advantage. Firms should not make available posts that users do not want to share, such as those marked private or limited to a group (although there are cases when they do, as described in Section 5.1). Geographic coverage can vary widely according to legal or political jurisdiction, and social media in many countries are censored (Levin et al. 2015). Also noteworthy from a research perspective, companies regularly change the information they share, often with little to no advance notice. They regularly shift the kinds of data relating to social networks, especially regarding the attributes attached to people and their relationships.

Data collection can be attenuated at the source by limiting what is collected. Users can often turn off the camera or microphone on their devices or limit when an application collects information. For crowdsourced data collection, Harvey (2013) distinguishes between data collected via an *opt-in* agreement – making it volunteered – as opposed to data collected via an *opt-out* provision that makes it contributed. Opt-in agreements provide more clarity and control over data collection, use, and potential reuse. In contrast, opt-out agreements are often so open-ended as to make it very difficult to control data collection and use, making it more challenging to assess the data's fitness for a given use. Either way, the savvy user can elect to limit data collection in potentially sensitive situations. It is unclear how many people have the time, knowledge, or intention to take these steps.

The tools and approaches of data science can have subtle biases. These go beyond passive sampling and representation issues found in social media and involve the active creation of bias in the real world. Social media firms use machine learning and other forms of artificial intelligence to guide people toward specific topics and forge relationships with other people. Social media are increasingly generated by *bots* (automated programs pretending to be people) or companies, making them even more remote from a random sample of most human populations. Furthermore, as exemplified by several high-profile scandals, a broad and diverse set of individuals and groups, including firms and governments, purposely spread misinformation (Del Vicario et al. 2016). Data science is both a consumer and creator of potentially biased information.

Other forms of bias can be inherent to core tools in data science. Notably, machine learning and other approaches are often trained against poorly specified sample data. This training can lead to situations where, for example, facial recognition algorithms do a poor job with women and people with darker skin tones because the learning data skews toward white and male (Buolamwini & Gebru 2018). Some social media users are more active than others or are seen as more authoritative for arbitrary reasons. These imbalances create positive feedback effects on who or what is reposted or retweeted, helped by data science algorithms designed to make material go viral (Liu & Zhao 2017). There is a large body of research on how people in both private and public contexts will craft an outward-facing image that is at odds with what they consider their true selves, moderating their information for audiences known and unknown. This situation is likely exacerbated in countries where social media is actively monitored and policed. Chapter 5 details data's various legal and policy challenges, including bias created and driven by data science approaches. As explored in Section 5.1.3 regarding surveillance, social media companies go beyond collecting data. They and other actors actively experiment with curating content to mold political and social thought.

The commercial nature of social media can create feedback effects in data collection and attendant data quality. People who fully participate in digital life have data profiles across systems that make it easier for new systems and analyses to latch onto the profiles and, by extension, the person associated with the profile. The "data rich" get richer because people with readily accessible and detailed profiles only become more so across a range of public and private spheres (Graham & Shelton 2013). Other people remain outside of the system – unknown or unmeasured, or perhaps even worse, confused with someone else. Chapter 5 examines how this visibility in the data realm can have a range of positive and negative impacts on people, places, and the quality of human-environment research.

Data science can draw on existing research on understanding and compensating for some data biases. The impact of bias depends on the questions being asked. Careful analysis of data derived from cell phones versus more traditional methods has demonstrated that these sources produce similar results for topics like human mobility (Wesolowski et al. 2013). More broadly, practitioners in most fields exercise sample control in ways that are too often underexplored in big data, such as physicists controlling and standardizing experimental designs or quantitative social scientists carefully designing and assessing their survey instruments (boyd & Crawford 2012). There is general agreement that machine learning and cognate methods are only as good as their underlying data. However, too much data science work still ignores or glosses over this fundamental issue.

2.3.3 Spatiotemporal Gaps

Human-environment data have spatial and temporal gaps. Some kinds of human-environment data are readily available, such as recent climate observations or online opinions about global warming. Other kinds of data are surprisingly sparse.

Human-environment data often have gaps in data collection frequency, such as years between censuses or weeks in remotely sensed imagery for many parts of the world. There are also spatial gaps in many forms of data across local to global scales. There is relatively little detailed information about many social and natural features over most of the globe, especially for the twentieth century and before.

The bulk of human-environment data is relatively recent or has temporal gaps. Important human-environmental dynamics like climate change and population growth have their origins in events decades and centuries ago, but we have little data on these and other phenomena. The information we do possess is often limited by how it was collected. For example, remotely sensed environmental data and national censuses offer a reasonable longitudinal extent, back to the 1970s for remotely sensed data and back decades or even centuries for some kinds of census data. Their temporal frequency can be limited. For example, many remote sensing sources capture data on weekly or longer timescales. The number of useful images for a given area is even lower when images are lost due to cloud cover or other atmospheric issues. If ground conditions do not align with available imagery, these limits are exacerbated. For example, it may be vital to capture the leaf-on and greening periods for a deciduous forest or a watercourse's spring flood.

Data science approaches may help bridge some temporal gaps. For example, there is a big push to develop gridded population data globally to help span periods not covered by population censuses. Many research groups are increasingly using machine learning and other data science approaches to model populations based on social media data, or older yet still relatively recent remotely sensed imagery (Leyk et al. 2019). There is also much interest in using data science approaches to assess climate change and its relationship with extreme weather events such as hurricanes, typhoons, and cyclones (Bongirwar 2020). However, there is disagreement on how accurate the underlying data sets are in the era before satellite-based remotely sensing (Chang & Guo 2007).

Human-environment data can also have spatial gaps. For example, environmental monitoring in many lower-income countries is limited to pockets of base stations focused on air quality, short-term carbon sinks, and source measurement (Kulmala 2018). Authoritative sources of information, such as Forest Resources Assessments of the United Nations Food and Agriculture Organization (FAO), can vary in quality. They are derived not from a single standard scientific measurement regime but from governmental statistics that vary significantly in completeness, timeliness, and accuracy within states and across continents (Grainger 2009). For example, some countries do not report any land change information. In contrast, others report information, but specific land use categories shift over time, making change analysis difficult. Permanent pastures or woodlands can be rolled into more generic categories, including nonarable land, nonpermanent cropland, or forest. These data challenges argue for broader scientific efforts to give human-environment researchers better data on a range of global phenomena. Many of the spatial gaps are related to the digital divides among people, groups, and regions (more on these divides in Section 5.2).

Spatial sampling frames used to collect data can lead to spatial data gaps. Consider the case of remotely sensed imagery, the use of which raises the issue of determining the most appropriate spatial unit of analysis. The pixel rarely maps onto any standard social science or environmental units of observation, such as households, settlements, fields, trees, or forest stands. Another related challenge with using regions to collect or aggregate data is that the regions rarely make sense for more than one or two kinds of analysis. Ecologists spend inordinate amounts of time carefully designing field campaigns that consider the spatial and temporal nature of the measured phenomena.

This task becomes much more difficult when more than a couple of ecological or biological dynamics are in play, common in human-environment research. There is often a scale mismatch because geographic and temporal scales do not align. Human data are often aggregated across subsets of the population, like health records or social media posts reported for arbitrary geographies like postal codes. In contrast, environmental data can cover broad extents that only partially map onto human jurisdictions because they correspond to ecologically coherent units like watersheds or biological communities (Fleming et al. 2014). Spatial samples that use these units may not necessarily map well onto other samples.

Sample frames contribute to spatial data gaps in social data as well. These data involve population counts and related measures such as age, income, and household characteristics. While they pertain to specific individuals and households, these data are often enumerated, analyzed, and reported for nations and subnational regions such as provinces, wards, or states. The tricky part is that the borders of these regions are arbitrary in most cases and often shift over time. National borders in many parts of the world are relatively new colonial artifacts drawn with a European-centric viewpoint, and often with little attention given to life on those lands.

Subnational boundaries can be very fluid since most countries reorder their census geographies over time. Some countries have engaged in the wholesale reordering of internal geographies, making it difficult to compare populations between years. Examples include Rwanda, where every census district was redrawn after the 2002 census, Poland after 1988–1990, and Chile after 1970, where many lines and districts were redrafted (Sula Sarkar, personal communication 2019).

These significant shifts are usually made for political or economic reasons or to make surveys better by bounding the population as it shrinks or grows. However, these changes also break the utility of the data for longitudinal research in ways that require careful expert analysis to remedy (Kugler & Fitch 2018). Section 4.1.3 further explores the effects of sampling regimes on data science.

2.3.4 Crowdsourcing and Citizen Science

Crowdsourced and citizen science are incredibly valuable to human-environment work, but they are not without problems. Perhaps the most significant is that crowdsourced data are usually not intentional samples. Data projects such as OpenStreetMap (OSM) or eBird are biased in how they tend to capture areas of broad interest, such as tourist sites and recreation areas. They can ignore areas on the margins of society, such as poor neighborhoods or difficult-to-reach environments, including wilderness areas (Haklay 2010). Similarly, these data tend to be developed by people with time and fast internet connections, which can be disproportionately in industrial nations or urban areas in many parts of the world, and involve many steps where errors can creep into the data (Degrossi et al. 2018). These sampling issues are not insurmountable, but they require care to address appropriately.

Related to sample-frame issues, crowdsourced data run into observer biases not typically found in research data. A significant form of bias is that citizen scientists are less inclined to record the absence of something. Recorded nonevents such as the absence of a species at a time and place can be very valuable for research. Collecting such data via crowdsourcing requires intentional framing of questions and survey instruments, such as asking participants how long they observed a given setting to establish periods where nothing was recorded (Cooper et al. 2012).

Sometimes big crowdsourced data can be too big. The Genghis Khan Tomb Project had thousands of people scour satellite and aerial photographs for archaeological sites looking for where the Mongol leader could be buried. This effort led to 2.3 million suggestions of where the tomb may exist, of which 53 were actual leads (Casana 2020, p. 595). Another example is eBird, introduced earlier, where the number of people reporting birds skyrocketed during the COVID-19 pandemic (Hochachka et al. 2021). Reasons vary but among them was that people were looking for new hobbies and had more flexibility in their schedules. Some challenges accompanied the boon of new data. The increase in observations made it difficult for scientists to determine the reason for an apparent uptick in bird prevalence. Was it a COVID-19–related factor, such as less traffic leading to more inviting habitats? Were more people observing the same number of birds and inadvertently inflating the counts? The spatial patterning of observations also changed in that urban areas were oversampled relative to rural areas as travel restrictions limited where people could go (Sánchez-Clavijo et al. 2021)

There are several ways to deal with error and uncertainty in citizen science. Successful projects strengthen relationships among individuals, citizen science programs, and scientific and governmental monitoring agencies to maintain data quality and expand spatiotemporal coverage (Poisson et al. 2019). There are technical solutions as well. For example, crowdsourcing spatial data can automate quality control and synthesis by imposing internally consistent and logical rules that follow general principles of geography and knowledge about place and space (Goodchild 2013). There are also promising ways to limit errors before they happen, such as imposing metadata standards or controlled vocabularies on measurements. Tools like the United Kingdom's National Biodiversity Network Record Cleaner help improve the quality of biological databases by using a range of logical rules to catch simple errors (e.g., impossible dates like February 31, incorrectly formatted spatial locations, or spelling errors) or to flag more advanced issues that require observer confirmation, such as finding animals outside of their normal seasonal ranges (Hassall et al. 2017). Computational approaches to automated error correction are explored more in Section 3.3.5, which examines smart computing.

Institutional and collaborative approaches are also successful in addressing errors in crowdsourced data. Technical error-checking methods are helpful for vetting data and help project experts better understand, work with, and accommodate citizen scientists (Boakes et al. 2016). Projects can also shuttle back and forth among data sources, such as moving between large-scale imagery and on-the-ground fieldwork by experts to balance big data with actionable findings (VanValkenburgh & Dufton 2020). Biological data can be verified with a mix of automated and manual approaches. These include more accurate ground-based sensors, automated validation tools within sensors, or post-collection verification workflows that rely on experts or several rounds of additional crowdsourcing (August et al. 2015). There is always room for developing and using training, such as offering online courses or asking volunteers to do tests to demonstrate competency (Crall et al. 2011). As with so much work in data science, the human and institutional elements are essential to successful workflows and outcomes.

Humans play another important but often hidden role in data science. Many data science approaches require algorithms to be calibrated against data sets that humans create. Typical tasks include having a human identify objects in a series of images or classify images or objects used to train artificial intelligence methods (more in Section 3.2 on data science tasks). Torney and others (2019) offer a fascinating case study using publicly sourced data to develop fast, accurate, and timely wildlife abundance estimates. They examine how to develop species abundance estimates from aerial photographs from a 2015 survey of wildebeest (a member of the antelope family) in Serengeti National Park, Tanzania. One longstanding approach is for humans to examine photographs and count the number of animals they see; in essence, each photograph survey acts as a sample from which to extrapolate abundance counts. Identifying and counting animals is labor-intensive, painstaking work. The authors had people identify and count animals in photographs via an online citizen science platform. These images were used as the training data for a neural network that learned to identify and count animals in various images (more on neural networks in Section 3.1). The authors note that this method, once trained, is faster and can be more accurate than the average human, but the methods depend on the data developed by citizen scientists.

Ongoing technological advances promise more citizen science contributions to human-environment data. Location-aware cell phones are becoming more potent as manufacturers add additional sensors for humidity, air pressure and quality, and other physical phenomena. Another advance is the Raspberry Shake, a low-cost seismograph based on the Raspberry PI computing system, which was designed to allow people to develop inexpensive computer hardware. These kits are no substitute for more exacting scientific instruments, but they are suitable for adding capacity to networks developed to study local and regional events (Anthony et al. 2018). Also compelling are data rescue efforts around the globe, where a mix of professionals, students, and others use data science tools over the Internet to transcribe and preserve climate and weather records (Mateus et al. 2021). Efforts such as the Old Weather project offers volunteers the opportunity to mark up and transcribe ship logs from the nineteenth and twentieth centuries, a task that requires human intervention because of the range in handwriting styles of the time, along with decoding descriptions of Arctic aurora activity (Wilkinson et al. 2016a). These and related projects point the way to an exciting future for the combination of data science, crowdsourcing, and citizen science.

2.3.5 *Integrating Administrative and Big Data*

Administrative and big data can complement each other. More valuable than the often-glib proposals to replace administrative data with big data is recognizing that high-quality administrative data can be used synergistically with big data. Some of the most compelling work on child hunger and development, for example, can only happen with the deep detail and careful curation offered by microdata on individuals and households (Balk et al. 2005). Another example is how political scientists combine administrative and big data to conduct human-environment research (Brady 2019). They examine how people's voting behavior changes after disasters or how political donations can affect voting on various issues that can bear on the environment, such as trade and industry. When big data is combined with existing surveys and census data, inferences about human choices and actions significantly improve.

Carefully collected and curated administrative data are hard to replace with other forms of data. Landefeld (2014) describes experiments replicating administrative data on a range of socioeconomic phenomena with big commercial data. While there are certainly some areas of complementarity, the match is rarely one-to-one. Even the most promising work using remotely sensed imagery and other sources creates data with 50–80 percent coverage (Johnson 2022). These data are beneficial in places without formal and complete social and environmental surveys but are not as good as these surveys. Each source has distinct advantages and disadvantages. Also necessary is transparency around data collection and estimation methods applied to data, rules for protecting data confidentiality, and proprietary methods used to create data.

Administrative and big data can be combined to offset their respective weaknesses. Two of the most significant drawbacks to big data are their errors and their often singular focus on a source or topic. Big administrative data help address these flaws, especially when multiple sources of administrative data are brought together, as when combining census data with social service or taxation information to learn more about families (O'Hara et al. 2017). National statistics organizations form a data community with expertise born of long practice in developing massive data sets that are useful because they are carefully collected, curated, and maintained. These organizations also arbitrate or assess the quality of messy or incomplete big data and act as trusted brokers among data sources.

Administrative data are high quality compared to many forms of big data. Extensive surveys and censuses generally involve an extraordinary amount of thought, care, and attention to designing questions that create socially and conceptually meaningful measures of the population (Shearmur 2010). In contrast, big data often have a found quality. Scholars sometimes have to gloss over how well (or not) a given variable captures the actual phenomena that it is supposed to measure (Graham & Shelton 2013). Consider a project like the National Historical Geographic Information System in the United States. It regularly ingests spatio-temporal census data from several sources, statistically integrates them across years, and geographically integrates locations and units to facilitate comparison across time (Manson et al. 2020b). The project offers robust metadata and web interfaces that facilitate easy user access and export to both data and metadata. It is used by thousands of scholars looking at human-environment topics. Of course, it takes time and expertise to create these data and a sustained effort to develop and maintain the necessary technical and social infrastructure (Noble et al. 2011).

Big data offers ways to fill the gaps in official statistics. This capacity holds especially true for nations with inadequate state apparatuses for collecting data. The ability to use sources including remotely sensed imagery and cell phone data could help lower-income countries, in particular, gather essential data. However, local officials would need to be involved in most steps. Data gathered at a remove by a third party such as a firm or nongovernmental organization can introduce new forms of error and uncertainty. Of course, as noted earlier, questions asked by ostensibly neutral public agencies for censuses and other formal surveys have their own biases. Government oversight is no guarantee of quality data, but it usually offers better prospects for transparency. Section 5.4 picks up some of these threads in the context of sustainable development.

Much of the initial hope and hype centered on big data to replace seemingly staid administrative data sources has been overblown, given the various challenges of data science. Similar conversations occur around discontinuing governmental observation systems such as remote sensing programs or weather monitoring networks and letting the private sector provide data. While using private data to examine a range of topics has advantages, and private firms are at the forefront of developing new data sources, there are dangers in having firms control what is now a common good in many places. Chapter 5 delves into the dangers of private firms controlling essential data, which include issues of cost and considerations around privacy, discrimination, digital divides, and informational sovereignty, among others.

2.4 Focus: Remote Sensing and Data Science

Remote sensing is crucial to human-environment research and has used data science tools for decades. As discussed in Section 2.2.4, remotely sensed satellite platforms and aerial vehicles gather imagery via sensors with spectral, spatial, and temporal characteristics. In addition to advances in capturing images are developments in analyzing them. Humans still interpret images by just looking at them. However, remote sensing has introduced many improvements over the years that draw on, and have enhanced, what are essentially data science approaches. These data and analyses have seen broad use in human-environment research.

Remote sensing is sometimes not considered an automatic candidate as a branch of data science, but even a cursory examination of its data and processing makes it clear. Remotely sensed imagery shares many characteristics that define big data, not just the basic "v"s but many others (Section 1.2). In terms of processing and analysis, remote sensing has been using methods central to data science, such as sophisticated algorithms to classify objects in images, for much longer than the data science community commonly recognizes. Remote sensing is also a foundational source of human-environment big data.

Remotely sensed imagery satisfies the three big data criteria of volume, velocity, and variety. They also meet others, including value, veracity, and exhaustivity via global coverage.

- In terms of volume, remotely sensed imagery constitutes one of the largest bodies of data globally, and Earth observation systems produce petabytes of data annually (Yang et al. 2017b). Data are growing exponentially as private and public entities launch many more satellites. For example, the firm Planet Labs (formerly Cosmogia) launched a constellation of over 300 satellites that collected millions of images every week over the past few years. This source joins hundreds of other drones, airplanes, and satellites that host sensors with various spatial, temporal, and spectral characteristics.
- Velocity is inherent to how data are beamed down to base stations in real-time or used to drive rapid workflows in human-environment research. It is increasingly clear to a host of organizations that real-time or near real-time analysis and visualizations are necessary for developing a range of products and decision-making, especially in fast-moving situations such as flood or wild fire response. Only a fraction of these data is processed in real-time, which points to the ongoing need for new ways to deal with data velocity (Section 3.3.5).
- Remotely sensed data can have great variety. They are highly heterogeneous due to being gathered by a broad array of sensors, stored in many different and often incompatible formats and data models, and created and processed via many different approaches. An extensive and heterogeneous collection of secondary data products developed from remotely sensed imagery is explored later.

Remote sensing has a long history of data capture and analysis and many ties to data science. Remote sensing was a civilian and military technology from the 1880s onward. Initially, balloons and even pigeons were used to carry still cameras into the sky. The world wars ushered in advances in using high-resolution cameras on airplanes, followed by high-altitude aircraft and satellites orbiting the Earth. From the 1970s onward, satellites and planes (and more recently, drones) were increasingly common, and images collected became broadly used in human-environment research (Mooney et al. 2013). As Section 2.2.4 described, remote sensing is constantly evolving and has become more powerful and nimble in meeting many data and analytical needs.

Remotely sensed imagery offers a crucial view into many human-environment issues. Remote sensing is a prime source of land use and land cover data, which is the linchpin for many analyses on agriculture, ecosystem services, hazards, and climate change (explored in Section 3.5.2). These data can distinguish among a dozen or more kinds of land cover. They span many scales and complement field-based observations in some places while also capturing global-scale human–environment interactions.

Given the variety of available land cover data sets, there is much ongoing work on automating their production and reconciling differences. One of the most pressing challenges is extracting fine-grained objects, such as picking out specific parcels, identifying cropping systems, or fine-scaled patterns in vegetation communities or human settlements. There is exciting work drawing on artificial intelligence to combine multiple data sets. One recent effort generated the footprints of over 125 million buildings in the United States from millions of aerial photographs and satellite images, processed using neural networks (Wallace et al. 2018). The applications of remotely sensed imagery can only grow.

A good deal of ecological and environmental work relies on the big data of remote sensing. Imagery is essential to most large land cover data sets, vegetation surveys, and environmental monitoring, such as mapping global- to continental-scale vegetation and vegetative productivity and disturbance such as active fires or flooding. Fine-scaled imagery is integrated with other kinds of data via data science approaches to track species incidence. For example, remote sensing is the basis for work using methods such as neural networks to identify individual animals and groups, including penguins and elephants (Duporge et al. 2020). This work steadily scales up the amount of data involved, including larger areas and longer periods of the kind being done with standard remote sensing approaches over large data sets (LaRue et al. 2015). Ground and aerial or space platforms are often integrated with these analyses. Animal tracking tags become smaller, allowing their use on a growing range of animals, and they can draw on cell phone networks or link to satellites (van der Wal et al. 2015). Drone technologies are also increasingly deployed in conservation science, although there are concerns that they can interfere with the plants and animals that they are observing (Ditmer et al. 2015).

Ecosystem services are benefits that accrue to humankind from the environment. These benefits vary in scope, but examples include food, water, energy, fiber, timber, and services such as carbon sequestration, biodiversity, and flood protection (Costanza et al. 1997). Remotely sensed imagery is essential to better understanding ecosystem services. For example, food provision as a function of agriculture, rangelands, fisheries, or agroforestry can be understood with remote sensing and data science modeling (Chen et al. 2018). Clean water is an essential ecosystem service necessary for food production, human well-being, and socioeconomic outcomes. Remote sensing can be coupled to machine learning and can measure, assess, and predict a range of hydrological phenomena, biodiversity, and sustainable wood harvesting (Willcock et al. 2018). Integrated systems such as the GOOS are designed to capture measures that are expensive to collect manually, such as river discharge, lake levels, soil moisture, permafrost, snow cover, and glacier outflows (Smith 2000). These and similar measures are essential to understanding human-environment dynamics, such as the role of snowpack in urban water supplies or periurban recreation (Crawford et al. 2013). Remote sensing and data science measure these and other ecosystem services.

Atmospheric remote sensing is a key to understanding a range of human-environment systems. Sensors measure a range of atmospheric constituents, including tropospheric nitrogen dioxide or fine particulate matter like soot and pollution via spectroscopy in specific spectra. These data may be used to develop and refine pollution models and effects on human well-being or land cover and crops (Blond et al. 2007). There is much work on developing sensors that can identify compounds, including carbon dioxide, ozone, and aerosols. However, such approaches often miss hundreds of other compounds of interest and cannot typically trace many processes or fluxes. Given the pressing nature of the climate crisis, there is particular interest in climate-forcing gases like carbon dioxide, methane, water, sulfur dioxide, and nitrous oxide (Lauvaux et al. 2022). Aerial remote sensing platforms have been designed and launched to sense these and other gases that can operate at moderate to high resolutions of several meters over areas of a few kilometers (Hulley et al. 2016). These sensors hold much promise, but more research is necessary to develop space-based platforms with higher-resolution data that can discriminate among different gases or surface materials. These will draw on spectral resolutions that differentiate surficial rocks and mineral classes, expand gas-detection capability, or generate more accurate land-surface temperature estimates (Hall et al. 2015).

Hazards exist at the confluence of human and natural systems and remotely sensed imagery is integral to analyzing them. Weather data collected by satellite sensors is vast and can be used to assess the extent of hazards such as flooding and wind damage, as well as secondary effects on agriculture and other human activities, such as longer-term impacts of hurricanes and flooding on crop yields. At scales beyond individual hazards, remote sensing is essential for monitoring the negative impacts of global environmental change, such as sea temperature and anomalies like El Niño alongside warming forces like volcanos and human-emitted greenhouse gases (DeFries 2008). Remotely sensed imagery is coupled with data science methods to monitor and assess threats like fire and deforestation, land cover degradation, and flood risk (Mojaddadi et al. 2017).

Remote sensing can measure socioeconomic and demographic characteristics. These social features are central to population–environment interactions and land-change science (Crews & Walsh 2009). Remotely sensed imagery can be combined with data science approaches to estimate population features related to dwellings or housing and socioeconomic characteristics such as poverty (Arribas-Bel et al. 2017). Imagery can also capture key climate change drivers, including the sources and consumption of fossil fuels and renewable energy resources, including solar, wind, geothermal, biomass, and hydropower generation (Buenemann et al. 2011). Satellite imagery can also capture facets of the demand side of the energy-climate equation. These data can measure and model socioeconomic drivers related to population, wealth, weather, and climate that drive energy demand (Kugler et al. 2019). The combination of remotely sensed imagery and data science opens new doors to understanding human-environment dynamics.

Ocean observation via remote sensing is increasingly essential to human-environment research. As Section 2.2.2 notes for Earth surface data, there are many fixed observation systems of the world's oceans, but most are close to shorelines. Remote sensing platforms are often oriented toward land observation, but some satellites have advanced sensors for oceanic and atmospheric observations. For example, the joint US National Oceanic and Atmospheric Administration (NOAA)/NASA Joint Polar Satellite System has microwave and infrared sensors. These measure moisture, temperature, and pressure. They also map ozone, cloud cover, radiant energy, weather, and terrestrial phenomena, including ice, fire, vegetation, and landforms (Moltmann et al. 2019). China, Japan, the European Union, India, and other nations and regions are also increasing capacity. Systems including Hyperion on the Earth Observer-1 platform or the Advanced Microwave Scanning Radiometer (AMSR-E) can measure properties of snow and ice along with soil wetness in addition to a host of complex variations in cloud cover, precipitation activity, and sea surface temperature (Parkinson et al. 2006). The Aquarius instrument uses passive microwave thermal sensors to measure global sea surface salinity that can be tied to climate change dynamics and sea-level rise, and so on to coastal flooding, impacts on freshwater, estuary health, and other ecosystem services.

There remain many gaps in the data for human-environment scholarship despite the incredible growth of data science and remote sensing. These needs are distributed across many research domains (Jurado Lozano & Regan 2018). In terms of the carbon cycle, especially components of the Earth system that govern carbon dioxide releases and sequestration, there is a need for better measures of vegetation, aboveground biomass, vegetation height, and relationships of land cover to fire and changes in moisture. There are related needs for high-resolution data on land surface energy balance and water cycles, soil moisture, albedo/reflectivity, surface temperature, subsurface soil temperature, and snow water equivalent. There are also narrow but critical needs to measure phenomena with relatively small spatial footprints relative to their climate impact, including volcanoes, dense urban areas, and coastlines. Finally, there are significant variations across nominally similar data sets, such as land cover or crop-cover data sets often derived from the same underlying sources (Pérez-Hoyos et al. 2017). Section 3.5.2 picks up some of these threads for land use and land cover data.

2.5 Conclusion

For all the legitimate excitement researchers should have for the vast expansion in the size and scope of data acquisition, this growth has led to a situation where collecting and storing these data is very difficult. A terrible irony of the data deluge is that we are mostly blind to the data we already possess, let alone able to use or combine them with other data (Borgman 2019). The hope for many scholars is that we are moving toward sensor networks that can integrate across scales, expand the capture of many kinds of measures, and highlight valuable information in the torrent of data. Data science enters the picture by providing the tools to help use many forms of data better and, per Chapter 3, offering methodologies that allow machines to collate and interpret data in ways that only humans could. Data science is a driver of – and increasingly fundamental approach to dealing with – human-environment data collected by a large and increasing array of methods.

3

Big Methods, Big Messes, Big Solutions

Many of the challenges with human-environment data are exacerbated by critical methodological shortcomings. Many core tasks of data science, such as data manipulation, machine learning, or artificial intelligence, are made more difficult by the complex nature of human-environment data. Data science and cognate fields are the loci of exciting research in addressing these methodological challenges. Information science on data lifecycles is being adapted to the needs of data science, including research in the methods of metadata, ontologies, and data provenance. Computer science and related fields adapt approaches like parallelism and distributed computing to work with big human-environment data. There is also ongoing work in cloud computing and high-performance computing to address the needs of complex spatiotemporal data sets. The private and public sectors invest heavily in smart computing, embedded processing, and the Internet of Things (IoT). An extraordinary amount of effort is dedicated to sharing data and workflows to support reproducibility in science. Finally, data science has advanced how it handles data that capture spatial and temporal patterns and processes.

3.1 Approaches

Data science joins many other fields in modeling relationships between input and output data. An outcome is a function of inputs, where the function represents a correlation, a classification, pattern matching, or a more sophisticated model. Of course, this way of thinking about data is not limited to data science, and many fields take the same approach. While big data and data science are often considered recent fields, they draw on decades of knowledge and methods used in various disciplines. Researchers in remote sensing, ecology, climate science, statistics, physics, and social sciences have long used methods that undergird data science. These methods include complex statistical tools, machine learning approaches, and core elements of artificial intelligence like neural networks that have been given new attention in data science and big data (Hindman 2015).

While some elements of data science have existed for decades, the field has spurred new advances by putting data front and center. It has encouraged developments that build on earlier work with artificial neural networks, adapt existing statistical approaches to massive data sets, and prioritize data-centric workflows (introduced in Section 1.3). Data science uses various tools to establish the relationships between inputs and outputs, including artificial intelligence, machine learning, and statistics. Some methods can be grouped into symbolic, connectionist, and hybrid approaches. Specific approaches like machine learning can be supervised or unsupervised, depending on the kinds of tasks being asked of the algorithms (Section 3.2).

This section focuses on prominent elements of data science to provide a basis for examining how this work relates to human-environment research. It aims to offer enough of an overview to provide essential context for the rest of the book but cannot do justice to the fast-growing data science literature, only some of which is cited. Similarly, there is no agreed-upon set of definitions of methods in data science, and this book does not try to establish one.

Artificial intelligence, machine learning, and statistics are interrelated approaches to developing models. Artificial intelligence is a sprawling field with a long history, with the modern version extending back to the 1940s (Turing 2009). Most definitions center on methods that allow a computer to mimic human cognition or abilities, such as problem-solving or learning within a larger environment. Many artificial intelligence approaches are new implementations of existing approaches. There is much value in positioning new technologies against the broader backdrop of the existing fields of cognitive science, computer science, psychology, philosophy, and others. This long history also helps explain why many of the terms we use in data science are used in different ways by various research fields (Wooldridge 2021). As Brady (2019) argues, computer scientists have traditionally focused on pattern recognition problems and predictive machine learning, while statisticians and quantitative social scientists were interested in explanation via well-specified and consistent models for hypothesis testing. Cutting-edge artificial intelligence scholarship can be traced back decades to work with knowledge graphs and artificial neural networks (Fogel 1994; Cowan 1995).

Machine learning is more narrowly defined as using a computational algorithm that gets better at solving a problem over time. Like other aspects of data science, there are disagreements on the relationship between artificial intelligence and machine learning, primarily on whether the latter is a subset of the former or whether machine learning is related to artificial intelligence only when methods approximate human cognition (Mohri et al. 2018). Machine learning has roots in the 1960s and uses a range of methods, some of which are decades old, like principal components analysis or statistical classification algorithms, while also using newer methods like artificial neural networks. Machine learning is often concerned with predictions, using data to make accurate guesses at a future state from current data (Agrawal et al. 2019).

Statistical methods are integral to machine learning. Most machine learning relies on statistical approaches, such as using measures of fit or similarity to guide learning. It is helpful to think of them as separate, given how statistics as a field was well established long before the advent of big data and data science. Most data science analysis involves descriptive statistics but quickly moves into inferential statistics and experimental design to draw rules or models from data. Many significant advances in data science involve adapting existing statistical methods to operate well over large data sets, being more amenable to continually updating over real-time data streams, or operating on local subsets of massive data sets (Chen & Lin 2014; L'Heureux et al. 2017). Various analytical techniques have been adapted for big data, including decision trees, regression forests, dimension reduction methods, and penalized regression. For example, climate change research uses many methods central to data science, such as pattern finding or outlier detection. It has updated their power over the years with developments in data science and big data (Faghmous & Kumar 2014b).

There has long been a divide within artificial intelligence between symbolic and connectionist approaches. While the data science literature sometimes glosses over this distinction, it is helpful in understanding how data science approaches work in different settings. The symbolic-connectionist divide is one way to separate artificial intelligence methods but is also relevant to modeling in general. Many symbolic methods can map onto logical or rule-based approaches well suited to implementation as computational algorithms. They are not artificial intelligence methods but are nonetheless valuable for developing models.

Symbolic approaches attempt to encode classical theories of mind or codify knowledge via structured grammars, logics, and rules that act on symbols that represent knowledge and facts. These approaches can work well but tend to be limited to narrow domains. Their use in medicine, for example, has long been of interest because the domain is narrow and much knowledge is already codified. While diagnosing ailments can be an art, there is also much science and measurement. Therefore, artificial intelligence in medicine has focused on encoding rules to mimic how an internist or physician would carefully work through branching pathways of diagnosis or treatment guided by threshold values in various measurements from blood panels, electrocardiograms, and other tests. Symbolic approaches commanded much attention for decades before fizzling out in the 1990s due to the systems being brittle and not working especially well in real-world contexts, which became part of a more extensive "artificial intelligence winter" (Haenlein & Kaplan 2019).

Connectionist approaches use computational analogs to brains and other biological forms of cognition. Connectionism gained traction in the 1990s by espousing that the way forward in artificial intelligence was not to focus on codifying knowledge in abstract terms via rules but instead by drawing on knowledge of how actual brains function. Artificial neural networks are computational analogs to neural tissue that learn from inputs to develop outputs and trace back to work in the 1940s on equating brains to electric circuits (McCulloch & Pitts 1943). Their use in human-environment research became common from the 1990s onward in spatial modeling of social, natural, and human-environmental systems (Wilkinson 1996).

Neural networks are composed of layers of nodes and connections that enable the network to learn. These layers have nodes and connections among nodes within and between layers (Figure 3.1). They have a set of input units, variables that are proxies for entities like pixels in an image or words in a text. These input nodes are linked to one or more so-called hidden layers composed of hidden nodes that approximate neurons in a biological brain. The deepest hidden layer of nodes connect to a layer of output nodes. The network is trained against a sample of inputs (often termed features) and outputs, such as image pixels in remotely sensed imagery and sample data with labels denoting land cover classification. The weights of individual links among nodes are modified over time to match inputs to correct outputs. Connections linking more-correct output nodes for any input are strengthened over time while less-correct linkages are weakened. Over time, the network develops better mappings of inputs to correct outputs. For example, in classifying remotely sensed imagery, a neural network will correctly discern the land cover category of individual pixels.

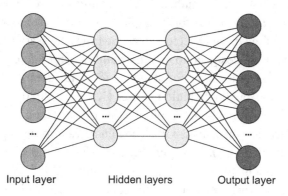

Input layer   Hidden layers   Output layer

Figure 3.1 Neural network structure with input, hidden, and output layers (Marcus 2018). Reprinted with permission of author.

Networks in data science are often more sophisticated than the example discussed here, but they have the same underlying principles.

Connectionism is the dominant artificial intelligence paradigm, fueled by increased processing power, a deluge of data, and successes in data science arenas. *Deep learning* is an advanced application of neural networks that uses multiple hidden layers, novel training methods, and a broader array of knowledge representations within the networks themselves. Deep learning and related methods deal with many big data challenges, such as heterogeneity, dirty data, or nonlinearity. A key advantage of deep learning is that it does well against many forms of raw data, while many machine learning approaches require relatively clean data sets as inputs. Deep learning approaches also tend to scale more easily because they can handle enormous data sets better than other machine learning methods. Much research is dedicated to making neural network approaches more sophisticated over time. Deep learning has proven apt for many forms of human-environment modeling. This work includes classification and pattern-matching problems (e.g., recognizing objects on the ground in remotely sensed imagery), predictions (e.g., developing hydrological forecasts), and developing synthetic observations or new patterns (e.g., tuning traffic flows in response to new information).

Hybrid approaches seek to combine different ways of modeling relationships in data. Some hybrid approaches result from renewed interest in offsetting the weaknesses of connectionist approaches by using symbolic ones. Mathematics, logic, and functions are efficient, easily extended to new situations, and require little data – characteristics missing from many connectionist approaches (Graves et al. 2016). For example, deep learning networks are expensive and time-consuming to train, require vast amounts of data, and are brittle with novel or high-velocity data, such as real-time observations in new settings (Chen & Liu 2016). Hybrid approaches draw on symbolic methods, such as using rules to preformat data or incorporating sophisticated mathematical abilities into nodes or layers. *Convolutional neural networks*, for example, explicitly encode known or anticipated subfeatures of the training set to reduce the time necessary to recognize them. For example, a network designed to identify buildings in remotely sensed imagery could be equipped with a library of standard footprint components or symbolic regression to help narrow down building sizes (Xie et al. 2020).

Researchers in many fields are exploring hybridity. These approaches include *embodied intelligence*, which means that cognition cannot be separated from the body or entity thinking, and *natural intelligence*, where a model mimics natural or biological phenomena in keeping with connectionism. *Social* or *situated intelligence* refers to situations in which multiple entities collectively find solutions to a problem, often concerning a broader environment. There is also interest in investing in connectionist approaches with findings from psychology on how actual brains work, such as creating stable long-term memory. Much research focuses on *reinforcement learning* of various flavors, where computers develop rules through trial and error (Sutton & Barto 2018). Finally, *ensemble learning* is where multiple machine learning approaches are used in concert. A typical application is where an analyst uses one kind of machine learning to take a pass on input data to produce a first-order result, followed by subsequent passes that narrow down the choice of algorithms (or parameters in a single set of algorithms) used to solve a problem (Dong et al. 2020).

Machine learning can be supervised or unsupervised. The heart of data science is associating input data with the correct outputs. Determining what constitutes correct outputs often involves *supervised learning*, where a human chooses cases from which a machine learns. A common form of supervised learning is classifying objects that have labels or descriptors, such as associating pixels in a remotely sensed image with labels of known ground cover such as forest or impervious surfaces. As explored in Section 5.2 on digital divides, there are entire industries where people classify imagery, text, or other forms of input to develop data–label pairs for machine learning. Seemingly mundane tasks like training computer algorithms against labeled images can ignore the reality that it is often low-paid and overworked humans who create the training images or engage in content moderation (Gray & Suri 2017). There is also much interest in crowdsourcing and citizen science as a source of quality data for deep learning and other approaches (as discussed in Section 2.2.7).

In contrast, *unsupervised learning* uses algorithms to sift through and discover patterns or structures with little human oversight. A common form of unsupervised learning is clustering, which has been used in many fields for decades to identify seemingly natural groupings or clusters in a data set. For example, researchers used a convolutional neural network with unsupervised learning to classify remotely sensed imagery to develop land use maps (Singleton et al. 2022). Even with unsupervised learning, there is often crucial human intervention, such as helping determine how many patterns or groups there should be when specifying data models (Yang et al. 2017a). Unsupervised learning is usually considered a more limited approach than supervised learning, but applications are increasing since they can often deal with messy and unstructured data in interesting ways.

3.2 Tasks

Data science approaches are applied to various tasks. Techniques used to do one kind of modeling are often repurposed to do other kinds, making it difficult to point to a single shared or canonical listing or hierarchy of approaches. Table 2 offers an overview of a half-dozen critical steps in the data science chain (Section 1.3). Chapter 2 looked at the first two steps of this chain: platform creation; and data generation and collection. Three more steps follow: preparation, representation, and transformation; aggregation, integration, and reduction; and exploration, analysis, interpretation, modeling, and visualization. This section grazes the surface of an exciting, fast-changing, and complicated aspect of data science (see Hernán et al. 2019).

Data munging is the process of moving data from its raw state to something that can be fed into later processes. Data are examined to discover any significant or unforeseen issues, such as large gaps in coverage or logical errors. Data are then transformed from raw data into data models or translated from one model to another, as when converting raw sensor readings stored in directories into an entity-oriented database. The data are then validated for data quality issues and sometimes enhanced by linking to other data (Boschetti & Massaron 2018). Data are cleaned to accommodate issues arising from error, uncertainty, and trustworthiness (Section 2.3.1 dives into bias, error, and uncertainty). Approaches range from examining data samples to more sophisticated exploratory analysis or quick passes with data mining to assess whether there are logical inconsistencies or gaps in the data.

Data standardization, matching, and modeling are important data science tasks. *Data standardization* involves developing systematic frameworks for measuring data quality and enhancing it via rules or guidelines, as noted in citizen science efforts. *Data matching* compares observations across different data sources to triangulate on seemingly more correct data values. Data consistency checks seek to validate data and processes to ensure that they do not change in unexplained ways over time. These checks include ensuring that data have not been corrupted and rerunning analyses on subsamples to ensure that the data are still suitable for the problem at hand. *Data modeling* involves storing or managing data in structures that enable subsequent analysis. Data models include entity-based, grid-based, and graph-based spatial data (as laid out in data basics in Section 2.1), but these are just a few generic kinds of dozens of possible models. Finally, all of these steps are usually connected via a *data pipeline*, the generic idea of moving data and resultant analyses from one place to another, where the output of one step acts as the input to the next.

Data mining, associations, and model building are catch-all terms for using computers to sift through large amounts of data to develop models and improve data quality. Data mining approaches vary. They tend to be machine learning or statistical methods, although the lines between the two can be blurry, and multiple methods can be tied together. Typical tasks include describing relationships among variables, such as establishing whether and how precipitation and temperature are codetermined as a function of elevation or other factors. A common approach in data mining is *association rule mining*, which finds close relationships between items in large data sets. Sometimes these are simple measures of association but they can also be more sophisticated rule-based equations and programs that draw on statistical or connectionist methods (Telikani et al. 2020). Many statistical methods have been adapted to varying degrees of success with big data sets, including most forms of regression modeling, nearest neighbor analysis, interpolation, kriging, colocation analysis, and hotspot and outlier analysis (Prasad et al. 2017).

Generalization is related closely to the process of developing associations from data. Indeed, a regression equation or set of equations is a form of generalizing data. More generically, generalization offers a simplification or description of inputs and outputs, such as extracting trends such as demonstrating how deforestation increases over time due to population pressure in a given location and decreases in other locations. Estimation and prediction are extensions of the search for associations via generalization. Estimation involves developing a valid model from data, and prediction is estimation focused on future states of a human-environment system (more in Section 4.1.2). A more narrow form of generalization is transfer learning. This machine learning methodology develops models for one problem or issue, termed the source domain. This model is then applied to a target domain, which saves the time required to develop a new model, or possibly solves a new problem for which existing models are insufficient (Weiss et al. 2016). Other work focuses on making algorithms for *lifelong learning*, an artificial intelligence program that can accommodate new situations (Chen & Liu 2016). Section 4.1 expands on the conceptual dimensions of making data science generalization less brittle in the face of a range of problems.

Classification involves categorizing or grouping input data. The distinction between supervised and unsupervised learning is especially germane here. Choosing appropriate cases or categories (often termed labels) and sample data for supervised learning takes care and attention. Membership can be considered probabilistic or discrete, either expressed as the likelihood of membership in multiple groups or belonging to a specific group. The latter is often expressed as a tree diagram or flowchart that helps interpretation. For example, a single pixel in a remotely sensed image is just a reflectance value. Turning it into a meaningful data point requires converting this reflectance into a category like "forest" or a value like "percent forest." There can be greater degrees of discrimination, like establishing the kind of forest (e.g., deciduous vs. coniferous) or how forest types may nest within each other (e.g., broad-leafed deciduous) within a taxonomy (a hierarchy of objects or phenomena).

There are many statistical approaches to classification that have been joined by machine learning and deep learning. *Cluster analysis* is a form of classification that does not use existing relationships. Instead, cluster analysis inductively (typically meaning unsupervised) determines similarity among objects. *Pattern matching* is another form of classification that overlaps with developing associations. It maps input data onto output data, as when sorting pixels into groups that correspond to categories like buildings or bird nesting habitats. Pattern matching commands much scholarly attention because it applies to many situations and can be hard to do well compared to human abilities. Development is ongoing, and methods are constantly improving (Bouveyron et al. 2019).

Dimensionality reduction is related to cluster analysis because it uses principal components analysis or random projection methods to reduce the number of data dimensions. Dimensionality reduction decreases the amount of data while preserving information from or the structure of relationships among variables, creating a more manageable number of variables or dimensions that reduce noise and accentuate meaningful data signals. With big data, this process can involve going from ten thousand variables to a few hundred or a few dozen, either by paring away the less useful ones or by creating composite measures that combine variables. For example, single-value decomposition models the values of observed variables as a function of a smaller set of unobserved variables. This approach reduces complex sets of variables into simple models of proxy variables in a way that is useful in capturing real-world human-environment dynamics, such as simplifying many variables in a flood-management data set down to a two-dimensional graph (Sood et al. 2018).

Visualization is a cross-cutting element of data science. Data science joins cartography and other visually oriented disciplines with hundreds of years of expertise in developing new ways of simplifying complex data and relationships. Data science increasingly deals with data sets so large as almost to defy human understanding. While scientists will always use numbers and measures of central tendency, there is much interest in better visualizations for understanding data. Indeed, there is also work in auditory and haptic feedback as well. The growth of augmented reality and virtual reality will undoubtedly extend to big data as researchers seek ways to spread the cognitive load of understanding data across the human sensory spectrum (McGregor & Bonnis 2017). Visualization is a valuable way to understand and use data.

Text, audio, and video analyses apply data science approaches to content from sources ranging from tweets to social media posts to documents, images, and videos (Maybury 2012). A range of approaches has been developed based on statistical analysis, data mining, and classification (Gandomi & Haider 2015). *Natural language processing* or *sentiment analysis* are designed to understand language and derive meaning or information from words drawn from many sources, from social media posts to governmental reports. For example, one may examine transcripts of interviews with farmers and other land users and other documents about agriculture to understand how people make land use decisions (Runck et al. 2019). The power and exactitude of sentiment analysis can be oversold in scholarly work and popular media (Puschmann & Powell 2018), but there are exciting challenges in extracting sentiments from various sources. These range from broad issues of conceptualizing categories or clusters of ideas to translating these concepts into measurements and then verifying how well relevant information is being extracted (Pang & Lee 2008). Methods for audio and video analysis have become vastly more sophisticated over the past decades, driven partly by the desire for their use in surveillance (Section 5.1.3). However, as we explore throughout this volume, these methods are also helpful for human-environment research.

3.3 Challenges and Solutions

Methodological challenges in data science rightly receive a good deal of attention. Some challenges involve data creation (Chapter 2) or how methodology relates to theory (Chapter 4). Many issues involve handling big data and making it more suited to different forms of analysis. Solutions for making data better center on the development and use of metadata and ontologies to facilitate data science. There is also a good deal of work, especially in computer science, on updating and expanding the underlying technology of databases. Related are parallelization challenges, spreading data handling and analysis across multiple computers. There is also much research on advancing the nature of computing, including high-performance computing, cloud computing, and other fundamental shifts in hardware and software. Finally, there is a good deal of work on making the basic building blocks of computing, like sensors and networks, more capable by imbuing them with internal processing capacity for onboard data handling and analysis.

3.3.1 Data Lifecycles from Provenance to Destruction

Data lifecycles are steps in developing and using data from initial conceptualization to ultimate transformation or destruction. Many data, intentionally or not, are part of a research cycle involving reusing and repurposing data that have "lives and after-lives" (Borgman 2019, p. 2). This lifecycle view of data gained currency in the 2000s as it became apparent that digital data were increasingly integral to scientific research and susceptible to damage, loss, and misuse. People can use centuries-old maps while digital atlases created a decade ago are gone or stuck on media that are impossible to read because we do not have the requisite software or hardware. The vital yet fragile nature of digital data argues for providing researchers with comprehensive support at all points before, during, and after its existence (Higgins 2008). The lifecycle approach can ensure that all of the critical stages are identified and planned and that necessary actions are implemented correctly (Bollacker 2010). This approach maintains the authenticity, reliability, integrity, and usability of digital material, which in turn promotes maximization of the investment in their creation.

The Digital Curation Centre (DCC) in the United Kingdom is a leader in digital information curation to improve education and research. The DCC Curation Lifecycle Model focuses on curating and preserving data via a detailed and realistic representation of the data lifecycle (Higgins 2008). It is notable in how it differentiates among three kinds of actions (full, sequential, and occasional) that map onto the frequency and intensity of any given task. The model centers on the data itself (Figure 3.2). Full lifecycle actions describe tasks that need to be borne in mind over other actions, including description and representation of information, preservation planning, community watch and participation (implying oversight by data stakeholders), and curation and preservation. In contrast, sequential actions relate to seven steps that more closely map onto the steps found in other life cycle models: conceptualizing; creating

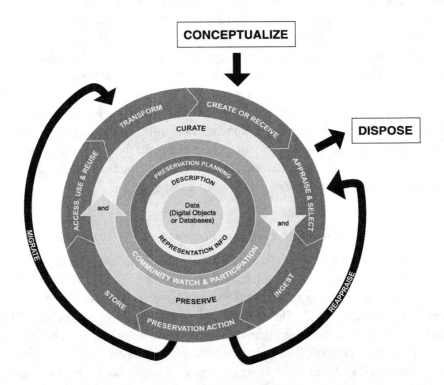

Figure 3.2 Digital Curation Centre Curation Lifecycle Model (Higgins 2008). Reprinted with permission of the Digital Curation Centre (www.dcc.ac.uk); CC-BY 4.0 (creativecommons.org/licenses/by/4.0/). Modified to render in grayscale.

or receiving (when created elsewhere) data; appraising and selecting these data; ingesting; preserving; storing; accessing, using, or reusing data (focusing primarily on data discovery and access by users); and finally transforming data. The model also identifies occasional actions that include reappraising, migrating media or format, and disposing of data or moving them to another custodian. This model has been applied successfully in the earth sciences. However, it requires well-trained staff, a commitment to ongoing consultation with stakeholders, and explicit recognition of the importance of lifecycle approaches (Bishop & Hank 2018).

A vital part of the data lifecycle is documentation. The Data Documentation Initiative (DDI) is championed by the Inter-university Consortium for Political and Social Research (ICPSR) in the United States in partnership with other organizations (Vardigan et al. 2008). The DDI focuses on developing and promulgating metadata specifications primarily oriented to social science data, but the model has been extended to many domains. It emphasizes the study concept and how it relates to data collection and processing and then to distribution, discovery, and analysis. This focus is helpful because it sheds light on an essential but underemphasized stage, namely, where a survey is being designed. What questions drive the work, and how do the questions and data follow? Importantly, it emphasizes the role of data archiving as a step between processing and distribution, and the repurposing of data after analysis. The DDI also specifies metadata requirements into five modules: study conception; data collection process; data encoding and logical structure; the subsequent physical structure of encoding; and archiving.

A significant effort in developing lifecycles for human-environment data is Data Observation Network for Earth (DataONE). This cyberinfrastructure project is designed to support data discovery, sharing, and access across a federation of existing data centers (Michener et al. 2012). The system gives environmental and biological researchers tools for key steps in the data life cycle, from collection to analysis. Like other successful projects, DataONE is less a fixed destination than an architecture that evolves via a process of working with stakeholder communities and engaging in usability analysis and assessment. It also provides a suite of tools that researchers find helpful, such as plug-ins and libraries for well-accepted programs like the R statistical package or the Zotero reference management program. Examples of successful application domains map onto nodes like the Oak Ridge National Laboratory Distributed Active Archive Center, which focuses on ecological and biogeochemical dynamics, or the Knowledge Network for Biocomplexity, which focuses on ecology and biodiversity. The DataONE lifecycle somewhat differs from others because it anticipates that new projects may skip steps, especially from preservation to analysis, if discovery and integration are not a primary goal.

3.3.2 Metadata, Ontologies, and Provenance

Metadata, ontologies, and provenance are integral to the lifecycles of human-environment data. Metadata, as explored in Section 2.1, are data about data. Related is a data *ontology*, a system that specifies semantic labels, relationships, and taxonomies, often domain-specific. *Provenance* is the record of where data originates and subsequent modifications. Metadata, ontologies, and provenance aid data science by facilitating automated data preparation, exploration, and analysis. Data science is, in turn, increasingly employed in creating and maintaining metadata, ontologies, and provenance information.

Metadata and ontologies facilitate the linking of data sets in data science workflows. A typical application is linking web documents using well-known approaches to implementation-independent encoding. One such method is extensible markup language (XML), which both humans and machines can read, although it is oriented toward the latter. Another is when people and systems use standard ways of linking to other documents and resources, which render it straightforward to access data sets or even recreate workflows (which become a form of provenance). For example, many human-environment data sets have a *digital object identifier* (DOI), a unique character string that references objects so that they may be easily found over the Internet.

The *Resource Description Framework* (RDF) is a more advanced form of metadata linking. It draws on standard conceptual modeling approaches used in computer programming, such as entity–relationship or object-oriented class diagrams. These tools describe relationships among digital resources, capturing key relationships, such as subject–predicate–object statements, termed triples. The subject of a triple denotes a digital resource, such as a file or variable. The predicate term denotes the relationship between the subject and an object, an attribute or characteristic, such as the file contents or variable measurement (McBride 2004).

Metadata and ontologies support successful data science of human-environment systems. Beyond linking documents or databases, metadata that specify essential spatial, temporal, and attribute characteristics make joining or reconciling differences between two data sets more straightforward. An ontology similarly aids structuring and guiding analyses by clarifying the meaning of data and relationships among entities and measures. For example, Morsy and others (2017) developed a metadata-driven framework that uses RDF triples to share and reuse hydrological models within the broader HydroShare modeling framework. HydroShare is a distributed system designed to enable researchers to store, manage, share, publish, and annotate data and models (Tarboton et al. 2014). The metadata-driven approach allows the connection of otherwise disparate models by defining and sharing new resource types for environmental models that define a range of objects (attributes) such as input data, model parameters, and outputs. Another compelling example is the KnowWhereGraph, which uses RDF and other metadata and ontology standards to bring together geographic information about human and environmental systems. This system allows people or computers to query locations or themes to see information about human-environment topics like the effect of a hurricane on agriculture or transportation (Janowicz et al. 2022).

Two approaches are generally employed in crafting ontologies, namely top-down and bottom-up. The *top-down approach* generally involves a group of experts or an agency establishing terms and relations that define objects for a given domain. These are usually formal ontologies, where terms are well defined and connected to allow for contextual information, logical inferences, and ways to structure and extend the taxonomy (Lauriault et al. 2007). Such an ontology can clarify relationships among components of a complex human-environment hazard system, for example, by naming and defining phenomena like floods, volcanos, and earthquakes, and how they relate to the material and energy flows of the Earth system (Masmoudi et al. 2020).

More recently, there has been a move toward *bottom-up* or hybrid approaches. These methods inductively derive ontologies using data science tools. Natural language processing or other approaches can extract semantic information from data, piecing together objects, categories, and relationships. There is much interest in the semantic web for large and heterogeneous environmental and human-environmental data sets. Data science techniques like machine learning can draw on existing ontologies or knowledge to help fill in gaps or correctly sort or identify new samples, such as new environmental chemical compounds (Hastings et al. 2021).

Ontologies are essential to dealing with the critical methodological challenge of brokering data across various systems. Massive cross-disciplinary systems like the Global Earth Observation System of Systems (GEOSS; introduced in Section 2.2.2) rely on a mix of harmonization (modifying data to use the same schemas) and brokering (mediating differences among different data sets) to facilitate the use of a variety of data sets (Nativi et al. 2015). There is much interest in linking and sharing metadata via portals and catalogs. The challenge is that often these systems will return thousands of results on data sets only tangentially related to the topic of interest.

There is a move to *application program interfaces* (APIs), software connections between computer programs that automate the provision of data or functionality. These combine with metadata to allow portals and catalogs to deliver more useful results when searching for data. The number of data-oriented APIs has increased dramatically over the past two decades. Janowicz and others describe how web-accessible APIs grew from about 100 in 2005 to 22,000 in 2019 (2020, p. 626). The increase in the number and power of APIs has helped make data more accessible.

Ontologies are valuable for data science but can be difficult to create and manage for many kinds of human and natural systems. Underlying knowledge can change and render existing ontologies obsolete, or critical incompatibilities can exist in neighboring fields, especially in epistemology and approach (Davies et al. 2006). Most fields have little impetus to fund the effort necessary to develop metadata, even within individual interdisciplinary projects with excellent funding and intentions (Borgman et al. 2012). Even newer and more automated processes have their drawbacks. For example, a shortcoming of using APIs for data linking and metadata provision is that the interface can be changed or the underlying data removed, as opposed to being able to download a set of data and archive it beyond the control of the data creator (Hogan 2018). There is hope that deep learning or other artificial intelligence methods can help automate ontology creation, and research is ongoing.

Provenance is an essential form of metadata. By tracking data from origination through subsequent modifications, provenance helps scholars draw on existing data sets, conduct analysis, and document any changes they make to their data (Musen et al. 2015). The ability to track every step in a workflow that starts with raw data and ends with publishing and subsequent responses is beneficial to the larger scientific enterprise. It can help others discover and share the research, identify any missteps along the way, help with replication of the analysis, and allow subsequent analyses to build on existing work quickly (more on study replication in Section 3.4.1). Data lifecycles increasingly rely on automated tools for provenance creation, freeing researchers from the time and effort of documenting their work. There is untapped potential to use newer data science methods and approaches, like database methods, distributed processing, or extending standards promulgated by organizations like the Open Geospatial Consortium (OGC) to ensure compatibility among approaches and systems in developing and using provenance information (Vitolo et al. 2015).

The big data of human-environment systems hold unique challenges for data lifecycles. Provenance, in particular, is often problematic. The importance of provenance is readily accepted in more traditional forms of information science, such as library science and archival research, and these fields have developed many well-honed practices. In contrast, provenance in big data is often underdeveloped. Much current work in big data is performed with relatively clean observations for which the provenance is straightforward because the data come from a single source, leaving aside the effort required to clean the data in the first place. These data's vast size and redundancy can often offset problems like missing values or errors, although there are limits even for relatively simple data sets. In contrast, human-environment data often are from singularly noisy platforms or different sources. These data are a provenance nightmare because they are difficult to source and are often manipulated before the researcher sees them. Many of our current approaches to provenance and related methods in big data are challenged by human-environment data, but data science is making promising inroads.

3.3.3 Computing, Parallelism, and Distribution

One way of thinking about many methodological challenges and solutions is whether computing can be scaled vertically or horizontally. *Vertical scaling* involves making a single computer or server faster with more or better processors, memory, or storage (Singh & Reddy 2014). Vertical scaling remains essential, but there is growing concern about whether individual pieces of hardware, such as processors or storage media, can continue to be made faster, better, or less expensive. For example, Moore's law posits that processors will double in power every eighteen to twenty-four months. However, many existing methods in chip manufacturing are approaching the limits of their ability to support this growth. Of course, history is littered with profoundly incorrect pronouncements about the impending doom of various technologies. Some breakthroughs in approach or material, such as quantum processing or new materials, may resuscitate Moore's law for processing. In the meantime, horizontal scaling is the primary focus of much data science.

Horizontal scaling refers to distributing computing workload across many computers or servers, where the speed or power of any one component is subordinate to adding new nodes to the network. Coupled to horizontal scaling is *parallelization*, or dividing a single computing task among computers, or distributed computing, where multiple machines take on tasks. Big data is primarily the domain of horizontal scaling, given that no single machine can be vertically scaled to handle many kinds of problems. Horizontal scaling can be inexpensive since it usually relies on commodity computing components. Of course, horizontal scaling is not without costs. Extra time and effort are needed to pass information among computers, which creates slowdowns due to translating information among many components. There are software challenges associated with dividing up a computational problem, solving it across processing units, and then piecing together a solution. These can include issues of *elasticity*, or how quickly data may be changed or updated, and *fault-tolerance*, or dealing with errors that creep into data and analysis as a function of being split up and moved around (Cuzzocrea et al. 2011).

Data storage and processing are essential to data science applications in human-environment research. Many advances focus on *distributed computing*, where data processing is spread among processing nodes or computers. Much of this work is in parallel processing, which usually refers to a narrower slice of distributed computing whereby a computational task is parceled out among processing nodes, and some data or information is shared in common (Ben-Nun & Hoefler 2019). Given how expensive it can be to shuttle data around, there is much research on managing how data and processes are spread over nodes. There is also interest in how various components fit together, such as tying computational procedures into an underlying file or database system. There is also ongoing research on reinventing the basics of managing data, such as security, access, and governance.

The parallelization process is often termed *split-apply-combine* because it operates by splitting data apart, analyzing or modifying them somehow, and combining the results of this analysis or modification. A canonical example of splitting a computational task in this way is MapReduce (Dean & Ghemawat 2008). *MapReduce* is the system in which data are divided among computing nodes. A computational operation is applied at each node, and then results are combined across nodes to provide a single answer. The terms map and reduce borrow from functional programming concepts, where *map* refers to applying a function at the local node and *reduce* refers to the process of collating answers from across nodes. The original MapReduce framework has been superseded in many real-world applications. However, the general term MapReduce is still used to denote the general concept of *decomposition*, or mapping data and operations across nodes and then summarizing results.

Problems solved decades ago for single computers have to be adapted for distributed computing. In addition to posing challenges in data storage, big data are computationally expensive. Machine learning approaches are like many other forms of computation in that the processing time or memory needs may increase as a square or cube function of the amount of data (in computational terms, $O(m^2)$ to $O(m^3)$, where m is the data set size) (L'Heureux et al. 2017). In these cases, doubling the amount of data will double the amount of processing time or memory and increase computing demands fourfold or ninefold. This work has several facets, focusing on parallelizing samples or using subsamples, applying divide and conquer approaches to big data sets, and online updating for streaming data that involves making calculations over data (Wang et al. 2016).

Another challenge for human-environment computation is that many standard algorithms assume data are *modular*, held in memory or stored in a single file. Much research in machine learning and data science is focused on speeding up or parallelizing processing, but this work is tightly tied to the storage model, and computer scientists seek new ways forward (Grolinger et al. 2014). Big data systems are built on a foundation of thousands upon thousands of individual components, including commodity-class hard drives and system processors and memory. These systems are built on the assumption that these components will eventually fail (Reed & Dongarra 2015). For example, many big data systems keep redundant backups and distribute multiple copies of a single data set among storage devices, which offers safety at the cost of redundancy and relatively slow transfer times (Yang & Huang 2013).

Computing approaches like MapReduce are implemented on top of a suitable file or database system. These systems manage the critical details of handling files across many computing nodes. The Hadoop distributed file system (HDFS, or usually just Hadoop) is among the most common file systems, as it is designed to process large, distributed data sets. There is ongoing work to improve these systems, but the basics have remained essentially unchanged. One or two central machines (actual or virtual ones) govern the split-apply-combine process and keep track of necessary metadata, such as where files are stored or their access permissions. Hadoop handles issues like redundancy by keeping multiple copies of data in case one or more nodes fail. Node failure is not uncommon, given that working on a problem may involve hundreds or thousands of computers. These systems lie at the heart of modern data science. They are joined by many others dealing with user interfaces workflow and scheduling, data integration, and other aspects of coordinating complex data and approaches (Eldawy & Mokbel 2016).

There is ongoing work to make data science methods faster and to better handle larger data sets. The Spark framework, for example, keeps much data in memory instead of on disk to reduce the time necessary to shuttle data around, at the cost of requiring a good deal of memory. It can also sit on top of Hadoop to handle massive data sets. Spark and the associated resilient distributed data set platform are gaining ground because they focus on performing analysis on these larger data sets in memory in a fault-tolerant way (Zaharia et al. 2012). In simple terms, this system will track the operations used to construct intermediate and final data sets; when one part of memory or storage fails, the system can reconstruct just the affected data. In handing complex spatiotemporal data common to human-environment research, resilient distributed data sets complement or even supplant MapReduce on file systems because they require less data replication and rely more on memory than on disk operations (Akkineni et al. 2016).

Relational databases have existed for decades and power many forms of research. Observations about a specific kind of data object, such as a hydrological station, are stored in rows for a series of attributes in columns in a data table. Different kinds of objects can be related to each other through linking tables, such as establishing how hydrological stations in one table are related to the water-courses in another table. Relational databases are the workhorse of human-environment research and can be integrated into data science workflows with some effort (Haynes et al. 2015). Relational databases are robust due to atomicity, consistency, isolation, and durability (ACID), designed to protect the data from accidents like a hard drive losing power during an operation (Haerder & Reuter 1983). *Atomicity* refers to how any operation is completed or not; there are no partial updates or incomplete operations. If an operation fails, it has to be redone. *Consistency* is the related idea that any transaction must be allowable under the rules of the database and preserve its validity and integrity. At the same time, *isolation* does the same for concurrent or cooccurring transactions. *Durability* means that once a transaction is complete, it stays within the table; for example, a power outage will not affect the proper recording of the transaction. Principles of ACID are essential for many kinds of real-world systems, but problems are posed by big data because they often involve multiple nodes and distributed data sets.

Many forms of data science analysis rely on alternatives to relational databases. For many data science tasks, losing some small portion of data to database or storage errors is acceptable as long as the errors are random. As a result, big data models may adhere to basically available, soft, eventually (BASE) principles instead of ACID ones (Pritchett 2008). The data system should be *basically available* in terms of being available to all users instead of being isolated and protected. The data can exist in a *soft* state where they may be changed, although they will *eventually* be made consistent as data are replicated over time across all nodes in the processing system. Adopting BASE principles has led researchers to develop alternatives to relational databases. One contender is the semistructured database, where data are stored in native formats such as text files or binary snippets but contain enough metadata so that they may be processed. These databases focus on having enough metadata to sort through digital objects intelligently, such as videos, documents, or social media posts. This sorting allows algorithms to make connections among objects, such as focusing on a subset of videos for real-time analysis in geospatial databases (Wu et al. 2017).

Human-environment data science has benefited from a rise in NoSQL approaches. Structured query language (SQL) is a standardized programming language for databases and is the lingua franca of relational databases. *NoSQL* languages are defined in opposition to SQL by not assuming there is a relational structure in a database. The "No" in NoSQL has gone from "non-SQL" to "not-only SQL" to connote that a NoSQL language can work on traditional relational databases and form SQL-compatible queries (Leavitt 2010). Of course, there is a great deal of hybridity, as with most things computational. NoSQL languages can operate over large relational databases, and relational databases can run parallelized (Haynes et al. 2015). Similarly, data science uses a range of more advanced data models integrated with analytical approaches. Several approaches use graphs, including probabilistic graphical models. These graphs can model an extensive collection of random variables as nodes and their presumed relationships and interactions as weighted links (or edges) between nodes. A successful human-environment project may require multiple database technologies, as in the case of a disaster analysis system that synthesizes hazards information from sources including social media, remote sensing, and web pages (Huang et al. 2017).

3.3.4 Cloud Computing and High-Performance Computing

Data science for human-environment research relies on evolving underlying hardware and software ecosystems. While a good deal of work relies on powerful supercomputing or in-house computing, many researchers and institutions are moving toward cloud-based computing systems. Standalone *high-performance computing* is faster on a per-core or per-machine basis and can be very helpful in doing many kinds of data science. Cloud-based systems implement parallelism to solve problems on a divide-and-conquer basis. Data science also benefits from newer processors tailored for specific data science tasks, especially deep learning and machine learning. Despite the many advantages of cloud-based computing, there remain important considerations around cost, expertise, and access with this approach and high-performance computing. The choice of one over the other for any given research task is not always clear.

Cloud computing supports much of the advanced work in data science, particularly deep learning. Many methodological problems, including the many demands imposed by parallelism, can be addressed by moving away from high-performance computing environments to cloud-based ones that hold a growing proportion of the world's data (Figure 3.3). In the words of its advocates, "modern cloud computing promises seemingly unlimited computational resources that can be custom configured, and seems to offer a powerful new venue for ambitious data-driven science" (Monajemi et al. 2019, p. 1).

While there is no doubt an element of hyperbole to this statement, it does capture the excitement felt by researchers who regularly wrestle with the quirks of high-performance computing environments. Google, Amazon, Microsoft, and other firms offer straightforward access to flexible, scalable, and robust generic computing nodes and a growing array of specialized ones designed to better deal with specific artificial intelligence jobs. The systems stand up and scale quickly compared to in-house high-performance computing. They obviate the need for space,

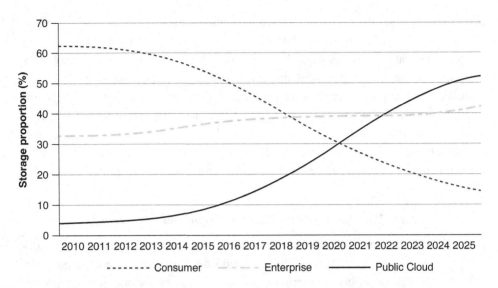

Figure 3.3 Where are big data stored? (Reinsel et al. 2018). Reprinted with permission from the International Data Corporation.

hardware, and expertise to manage in-house computers. The advantages of cloud computing are compelling, especially in how these services offer on-demand computing services and elasticity in terms of increasing or decreasing capacity (Yang et al. 2017b).

Cloud-based computing is not without problems, and in-house high-performance computing offers advantages to researchers. Cloud-based computing can be costly for larger-scale experiments or tasks like deep learning. Companies like Amazon and Microsoft give researchers ready access to computing in the low double-digit petaFLOPS range. However, even moderately extensive training can still be expensive, and very complex jobs can cost millions of dollars (Silver et al. 2017). In contrast, institutions subsidize high-performance computing environments so that students and others without research budgets can get time with a computing cluster. Accessing cloud computing can also require expertise in sophisticated software stacks or sets of programs and workflows that work together on a cloud platform. Internal high-performance computing often has support staff available to help researchers implement their projects. Cloud-computing systems can impose unanticipated costs when transferring large amounts of data over the Internet. In contrast, internal systems often have dedicated high-speed connections or offer the option of physically moving data in external hard drives or other storage media to a nearby computing center.

Even relatively simple cloud-based systems can be expensive. In the United States, NASA is moving some of its storage and access to commercial cloud-based systems.

The organization projects spending up to US$30 million per year for data storage and user access by 2025. These costs are driven by how the amount of data is projected to increase exponentially over time, from about 32 petabytes to approximately 247 petabytes by 2025, as new science missions begin producing data (NASA 2020, p. 12). NASA notes that a significant source of financial uncertainty is the high additional costs incurred when people download data from commercial cloud services. It is not just the cost of storage that matters but how much data users extract from the system. Limiting these egress charges will likely require developing the capacity for users to analyze these data on the system itself instead of downloading data to local computers or moving the data to another online or cloud-based environment. In the worst-case scenario, the amount of data that can be downloaded would be limited, including capping downloads entirely for periods. Finally, it may be too expensive to switch to another vendor because the costs of actually moving the data are prohibitive, meaning the data are locked-in to a specific firm.

Deep learning is a big driver of specialized computing hardware in cloud-based and high-performance computing. Over the last decade, the amount of processing power needed for the specific challenge of training neural networks has doubled every three or four months after relatively slower growth from 1950 to 2012 (Amodei et al. 2019). Researchers are pursuing much larger tasks than before but at the cost of correspondingly giant leaps in computational intensity. Part of the answer to meet this demand is the use of *graphics processing units* (GPUs), specialized chipsets with onboard memory designed to handle floating-point operations and other functions associated with graphics processing. Before 2012, dedicated GPUs for machine learning were rare or experimental, and processing times decreased on the order of magnitude of Moore's law. Over time, GPUs became more common and sparked the development of attendant methods in training, parallelization, and storage. Deep learning and other artificial intelligence methods also increasingly rely on developing and using specialized collections of computer chips. For example, Google developed tensor processing units to serve its TensorFlow framework, an open-source programming library for machine learning, particularly neural networks (Cass 2019). This blend of commodity-level processors and specialized hardware will likely continue to characterize work in data science for a while.

3.3.5 Smart Computing Making Sense of Stupid Data

Increased use of smart computing offers to mop up some of the messiness of human-environment data. Much of the labor in handing big human-environment data is during the *extract, transform, and load* (ETL) stage of data handling. ETL is the generic term for extracting data from sources, transforming them to match a given data model, and loading them into that model or a data warehouse (Miller & Mork 2013). The data are then available for subsequent processing or analysis. One way to reduce the demands of handling data is to reduce the amount of raw data and do more processing and analysis on sensors and other devices that collect data. For example, there is interest in *stream processing*, instrumentation, methods that take in large amounts of real-time data and convert it on the fly for analysis. Much research is being done on developing and using the IoT and other downstream computing and sensing modes.

Smart computing for human-environment research can involve better algorithms for compressing data or preprocessing raw data into a more refined format. Signal processing has been a challenge from the earliest days of computer processing. The profusion of sensors across many human and environmental systems poses a challenge and opportunity. While much focus is on images – cell phone pictures, remotely sensed imagery, or drone footage – there has been similar growth in other kinds of data. This work is necessary considering how much of the data we collect now cannot even be transmitted to an eventual end-user, let alone used. Ongoing research is needed on how to compress, sample, or model data as it is collected to reduce its size at its source (Baraniuk 2011).

Although not in everyday use, the term *extreme big data* denotes the next generation of advances in computing where supercomputing and cloud computing tools will be joined (Matsuoka et al. 2014). This coupling will require advances in memory, especially *nonvolatile memory* (where the information is not transient or lost when power is removed from the storage medium) and *processor-in-memory* approaches (where the computing and memory are on the same chip). Also necessary are computational methods that can better accommodate human-environment data's noisy, multigenerational, multisource nature. These approaches model how the data were collected to offset the negative impacts of sampling bias, errors, and multiple spatial and temporal resolutions to deliver data that accurately represents the real-world phenomenon of interest.

Advances in compression or within-sensor data selection are not costless. Many issues around scientific replication are exacerbated by how big data are processed. Many data science approaches are *lossy* because they do not keep all data or can accommodate losing a small percentage of observations to accidental deletion, noise, or incorrect formatting. However, discarding the raw observations may introduce a systematic bias unless it is done well. As explored in Section 2.2 on data sources, the process of capturing new measurements is where some pruning may occur via judicious use of logical rules. Some of these rules are straightforward, like only accepting field observations that have happened in a specific window to guard against incorrectly encoding dates or times.

Others are a little more fraught. Christie (2004) offers a fascinating exploration of missteps and missed opportunities in discovering the atmospheric ozone hole in the early to mid-1980s. In essence, the hole could have been discovered earlier. However, how satellite data were automatically analyzed and flagged led to a delay in researchers appreciating the magnitude of change in observations. Additionally, the ground-based instrumentation that could have provided additional evidence was missing, malfunctioning, or miscalibrated. Smart computing offers promise, but it is like any other approach in that there will be steps and missteps.

Embedded computing is another way forward in streamlining flows of human-environment data. *Embedded computing* is when devices like sensors are given their own computational abilities, such as summarizing readings instead of simply passing on reams of raw data or using machine learning to select "interesting" data. For example, a temperature sensor in a stream may only report the average daily temperature unless it senses a rapid change that could correspond to a cold snap or a dump of industrial wastewater. At this point, it reports data every second.

Related is the idea that sensors could be granted growing degrees of intelligence. For example, nest boxes are purpose-built places that entice birds to nest in them so that researchers can observe their behavior. Modern incarnations of these boxes can house one or more cameras that periodically take an image. These images may be fed into algorithms that determine whether the box has an occupant and what they are doing, including building a nest or laying an egg, thereby saving the time and effort of human observers needed to interpret recorded bird behavior (Cuff et al. 2008). This application is just one of many examples of the interest in moving core data science tools, such as machine learning and pattern recognition, downstream into sensors and physical optical and mechanical devices themselves (Wright et al. 2022).

The IoT is an expression of embedded computing that holds promise as a way to provide data and analysis of many human-environment systems. As introduced in Section 1.4, the IoT is the body of inexpensive and internet-connected devices such as cameras that record vast amounts of potentially useful information (Kortuem et al. 2010). Adding to the value of embedded computing and the IoT is *automated semantic derivation*, the ability of computers to translate raw images and data into information on real-world entities, a job now typically reserved for human analysts. Building new sensors with onboard computing reduces data transfer and storage needs outside a local network or even outside the sensor. A sufficiently complex and ubiquitous system of internet-connected devices is the basis for what may be termed a digital Earth nervous system that offers real-time awareness of hazards and other significant events (de Longueville et al. 2010). This nervous system then becomes the basis for a task such as monitoring air quality. Systems are set to trigger alerts when these readings surpass threshold measures associated with

anomalies like spikes in nitrogen dioxide associated with traffic jams or unanticipated power plant emissions (Trilles et al. 2017).

Using the IoT for research is not without challenges. Most inexpensive or
common environmental sensor devices and networks do not have metadata or
readily discoverable interfaces or APIs. Related is the poor interoperability
among components or difficulties linking them to dissemination mechanisms.
These challenges are exacerbated by firms using proprietary approaches for many
aspects of their sensors, including logging measurements or updating software or
firmware (Kotsev et al. 2015). Better harnessing the IoT for environmental work
will require plug-and-play sensor deployment, platform independence, simple
access to sensor data, and adherence to standards in order to facilitate data integration. Computing that resides at least in part in devices as part of the IoT is termed
edge computing. These machines lie on the boundary between low-powered
devices in the IoT and more powerful networks and computing systems. There
are several different terms for this work, including a variant of cloud computing
termed *fog computing*, as in cloud computing, but closer to the ground in terms of
devices (for a comprehensive overview, see de Donno et al. 2019).

Aspects of the IoT that once seemed like science fiction are becoming a reality. One
vision is a study site or region coated with *smart dust*, tiny computers the size of a grain
of sand that possess sensors and can communicate with other dust motes and larger
base stations that collect and collate sensor readings (Warneke et al. 2001; Ilyas &
Mahgoub 2018). Much more common and prosaic is a large array of inexpensive
sensors that can measure simple to complex phenomena. Some sensors can track a host
of gaseous and liquid compounds, while others can measure electrical resistivity to
measure metal corrosion or fluid salinity. Projects like the International Cooperation
for Animal Research Using Space (ICARUS) seek to reduce the size of transmitters
and make it easier to track them from orbit to make it possible to track smaller animals
over more extensive ranges. Finally, there is the ongoing development of sensing
devices that derive energy from their environment via solar arrays or piezoelectric cells
that can create electricity from motion induced by waves or the wind. Freeing sensors
from the need to have long-lasting batteries or permanent connections to the power grid
would eliminate a significant source of expense in building them or the need to
regularly maintain or service them (Paradiso & Starner 2005).

3.4 Sharing Data, Code, and Workflows to Advance Science

Reproducibility is vitally important to data science and human-environment research.
Reproducibility requires data, reusable code, and documented workflows that allow
the reproduction of scientific results. There is increased interest in data sharing in

science, but any conversation about data must eventually extend to model sharing. Data and models are increasingly inseparable, and it is necessary to extend data policies to models that create or examine these data. Tools and infrastructure that help researchers use and share common workflows offer one of the most promising ways forward for reproducibility in the data science of human-environment systems.

3.4.1 Repeatability, Replicability, and Reproducibility

The Association for Computing Machinery (ACM) offers "the three Rs" of sharing models: repeatability, replicability, and reproducibility (described in Plesser 2018). Repeatability is a low bar to hurdle because it only entails that the same team and same experimental setup should be able to run an experiment and get the same results. The same team finding the same results in the same setting is the bare minimum for science. Replicability exists when a different team with the same experimental setup can get the same results as the original team. Finally, reproducibility involves a different team using a different experimental setup to get the same results. Applying this schema to data science is superficially straightforward. Repeatability ensures that the researcher can reliably repeat their own computation, while replicability implies that a different group will obtain the same result using the original data and computing systems. Full reproducibility means that a different group can obtain the same result using data and computing that they develop independently of the original data and software environment.

Some commonly used reproducibility strategies are less suited to data science than other fields. The field of metascience is a site of much research on repeatability, replicability, and reproducibility (Munafò et al. 2017). This work is designed to help traditional forms of analysis, but some of the work applies to data science. Some strategies espoused by reproducibility proponents are at odds with data mining and other data science approaches. For example, preregistration of study protocols or expected findings is appropriate for randomized controlled trials because the data, methods, and hypotheses can be fixed ahead of time. The same cannot be said for many data science approaches because researchers will employ various approaches to tease out findings inductively. Questionable practices such as *p-hacking*, selectively rerunning models to find seemingly significant results, or hypothesizing after the results are known (*HARKing*) are serious issues across science (Simmons et al. 2011), and data scientists must be aware of them.

Successful replicability for human-environment research is challenged by spatial heterogeneity and spatial dependence in data and analysis. In simple terms, places are different, and space matters. Nichols and others (2021) argue for replacing or at least augmenting strict notions of replicability with more intentional sequences of studies directed at accumulating evidence over competing hypotheses. Goodchild and Li (2021) extend this idea to the notion of weak replicability as intentionally shuttling between the search for general principles and accepting that all places are unique. Weak replicability allows for how "the model specification might be generalizable, for example, but the model's parameters might be allowed to vary spatially and temporally" (p. 6). Similar conversations are happening in ecology, one of the leading fields for data sharing. Options include exact replication in studies for well-defined biological phenomena in limited locales, conceptual replication for species or systems, and quasi-replication for general biological phenomena across species or systems (Nakagawa & Parker 2015).

Uncertainty in most spatial data that undergird human-environment research adds to challenges in reproducible science. In particular, geographic analysis is complicated by uncertainty in conceptualization, measurement, analysis, and communication (Kedron et al. 2021). The value of this framing goes beyond reproducibility (Figure 3.4). These uncertainties underlie the entire scientific research cycle from initial hypothesis generation to publication. Per Section 3.3.1 on data lifecycles, this cycle should also capture others' reuse and extensions of these data. The number of decisions (and attendant possibilities for propagating the effects of uncertainty) increases with each research stage, from conceptual uncertainties to communication uncertainties. Human-environment researchers need to account for these facets of uncertainty as they embrace data science.

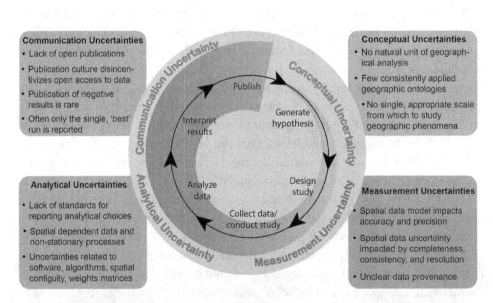

Figure 3.4 Sources of uncertainty in spatial data (Kedron et al. 2021). Reprinted with permission from John Wiley & Sons.

3.4.2 Data Sharing

The FAIR principles for research data are findability, accessibility, interoperability, and reusability. *Findability* rests on the notion that data should be described, identified, and registered or indexed transparently. *Accessibility* implies that data have readily available metadata, allowing researchers to gain access through a well-defined procedure. *Interoperability*, introduced in Section 2.1, means that data and attendant metadata are expressed in well-known or understood data models that allow spatial, temporal, or attribute characteristics to be compared and reconciled. *Reusability* encapsulates the other elements and extends the general idea that data are described sufficiently to allow for use in situations beyond the one for which the data were initially collected (Wilkinson et al. 2016b).

Like other aspects of big data, the extent to which researchers can enact FAIR principles depends on methodological and institutional support. Making data available and accessible works when there are shared standards around metadata and discoverability, which may war with privacy issues and appropriate use of complex data sets (Boeckhout et al. 2018). Documenting and sharing data (or code and workflows) in repositories and persistent identifiers such as DOIs helps establish provenance and subsequent changes. Of course, data can have an identifier and yet not necessarily be the exact data used in the model unless care is given to ensuring that object identifiers are kept up to date or assigned to specific data draws instead of generic data sets. Land cover data sets are sometimes reprocessed with better algorithms, for example, after published analyses, so care is necessary to ensure that the attendant identifiers are updated. There is also a need for better access and visualization tools. These can be very expensive and require in-house expertise in programming, networks, and web design. Some fields started early in developing shared repositories, including genetics and genomics. These fields often benefited from having domain scientists invested in learning the basics of sharing data from the earliest steps of data collection through to later stages of use and dissemination.

The number of journals that expect or require data to be shared is still low. Consider the journals managed by the publisher Springer, which have marked differences amongst fields (Klemens 2021). Interestingly from a human-environment research perspective, ecology and environment are among the highest-ranking compared to fields like psychology or medicine, where reproducibility has been a topic of much discussion (Figure 3.5). However, the standard for sharing often is that the author states that the data shall be made available on request, but tests of these policies show that less than half of these requests are honored. In the words of one article title, "Ecological data should not be so hard to find and reuse" (Poisot et al. 2019). Part of the challenge is that many fields do not have standardized data formats, which adds a hurdle to sharing before other actions can be taken.

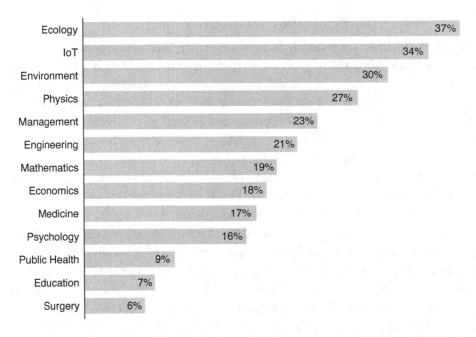

Figure 3.5 Percentage of journals with availability requirements (Klemens 2021). Reprinted with permission of author.

Poor incentives are a significant challenge in data sharing. As referenced earlier, ecology is a field that seems to be more invested in sharing, but problems remain. Incentives are hard to craft. They generally fall into career development, including tying data sharing to tenure and promotion or salary increases, and funding, such as external granting activity or internal infrastructural support (Organisation for Economic Co-operation and Development [OECD] 2007). In many institutions, neither career development nor funding processes focus on data sharing compared to other activities. For example, it seems that few funding agencies actively monitor data sharing or tie future funding to past data sharing, and institutions prioritize standard activities like publishing or winning awards as a pathway to promotion or salary increases. However, there is growing recognition of the importance of less formal incentives, such as badges and external markers on publications that vouch that data sets are in an open repository (Schöpfel & Azeroual 2021). To be successful, however, funding agencies and other organizations should have teeth to their data sharing mandates, such as limiting repeat funding in the absence of proof of data sharing in previous funded work, alongside providing technological and institutional support (Suhr et al. 2020).

3.4.3 Model Sharing

Model sharing is as important as data sharing but arguably less developed for most human-environment research. While progress is being made in sharing fundamental data sets and opening up model code, many human-environment fields have yet to reach a bare minimum in many respects (NASEM 2019). There are several steps scholarly communities can take to establish a robust ecosystem of practices and institutions around sharing data, coding, and workflows. These steps include having better standards for data sharing, readable and sharable code, automated workflows, centralized repositories, and computing systems that bring together all the previous steps. Reasons for model sharing are often the same reasons advocated for data sharing. Among these are, in short, to reproduce or verify research, to make the results of publicly funded research available to the public, to enable others to ask new questions of extant data, and to advance the state of research and innovation (Borgman et al. 2012).

One factor that likely holds back many fields from developing shared models is that they have de facto "sharing" because their models are widespread or seemingly straightforward in ways that occlude their underlying complexity. Regression-based models, for example, can be specified as a mathematical equation or through recourse to a generic description of a well-understood model, such as ordinary least squares or logistic regression. Why share model code when employing an almost universally used and understood approach? There are many cases where researchers have failed to replicate seemingly straightforward analyses, even where both the data and model formulations have been freely shared. Causes for this failure vary but can range from apparent problems, such as unreported variable transformations, to more subtle issues, such as different statistical packages (or versions thereof) having slight variations in implementing seemingly standard approaches (Stodden et al. 2018).

While model sharing and shared codebases are positive for reproducibility, there is the danger of commonly held tools serving as single points of failure. Chemistry is a reasonably mature field in terms of data sharing, and yet in 2019, it was determined that many hundreds of papers might be incorrect due to an error. A research team discovered a programming issue in programs commonly used in computational chemistry analysis. These "Willoughby–Hoye" scripts (named for their authors) would deliver different results when run in a different operating system environment. Many published chemistry studies could be negatively affected if these differences result in analytical errors (Bhandari Neupane et al. 2019). Even if fewer studies than expected have to be redone due to this issue, the situation highlights where failure may occur. Then again, the fact that the scripts were commonly available via open and shared code makes it more likely that people can find and correct errors.

Model sharing has a lot to do with the larger research culture. Some scholarly domains see model sharing as essential to scientific discovery. In contrast, others see models in proprietary terms, where data may be shared but models are protected for as long as possible (Nosek et al. 2015). Some research areas, like climate modeling, have a suite of commonly held models that are shared in the sense of being readily accessible (if not necessarily understandable to someone without advanced degrees in computer science, mathematics, or physics). Others, like integrated assessment, have shorter histories of sharing a relatively small number of well-accepted models (Kitzes et al. 2017).

Agent-based modeling of human-environment dynamics has a strong record of model sharing. Janssen (2017) advocates for policies that require model archiving using agent-based modeling for any paper published in *Ecology and Society* and other venues. Agent-based modeling also has modeling documentation formats. The overview, design concepts, and details (ODD) schema is becoming a de facto standard for many journals (Grimm et al. 2010). This approach works, in part, because the journal specifies how the data and models should be shared and that there is now a broader community-wide conversation around model validation (Manson 2007).

There is a deep need for better practices for making shared code usable once shared. Placing model code in a repository like GitHub or sharing R-code is a good first step beyond what many scholars do. However, usable code relies on the goodwill and time of researchers who take the steps necessary to clean up the code, document it properly, and then place it in an accessible repository. Sharing code libraries or enhancing base libraries that are widely used is a way to contribute to specific research projects and the broader field. An ever-present challenge is that underlying model languages and systems change rapidly. Operating systems become obsolete and can threaten to strand programs that rely on them. Generic computing languages and model-specific languages rise and fall over time, as seen with many agent-based modeling languages (Manson et al. 2020a). There are many other related challenges in developing, maintaining, and running software. For example, the British Ecological Society reported on reproducible code in ecology and evolution, laying out basics such as naming and organizing files or using scripts instead of handling raw data (Cooper et al. 2017). It needed to cover many basics because these methods and approaches are new to many scientists.

Beyond software issues lies the gnarly mess of ontology and knowledge representation. Modeling sharing schemes like CellML in cellular biology succeed because the domain is highly specialized and agrees on core concepts (Clerx et al. 2020). In contrast, modeling well-known diseases or conditions like malaria or hypervitaminosis involves representing various social and environmental conditions. Ontologies in biology are well developed, especially for model organisms like the fruit fly, which is a linchpin for many studies. Nonetheless, "despite the widespread recognition that model organism databases are among the best sources of Big Data within biology, many biologists are suspicious of them, principally as a result of their mistrust of the categories under which data are classified and distributed" (Leonelli 2014, p. 6).

Similarly, efforts to standardize agent-based models also illustrate various needs for advancing model sharing. For example, the ODD standard may be too abstract for many modelers. It is detailed but may not provide enough specificity for model replication. Extensions are proposed by researchers who feel that the generic formulation is not defined enough in how it handles decision-making or networks (Müller et al. 2013). While this confers flexibility and specificity, the single standard risks fracturing into many substandards after a few years following creation. There is also room for updates, as with the second update to ODD, offering advice on structuring complex ODDs, and addressing issues such as model rationale and evaluation (Grimm et al. 2020).

Hydrology is another example of a field that has made much headway in model sharing and yet exemplifies the many unanswered challenges. The provocatively titled article "Most computational hydrology is not reproducible, so is it really science?" reveals how a lack of transparent data and workflows imperils much work in water and hydrology (Hutton et al. 2016). Hope is offered by systems like HydroShare because it embraces the idea that data and models are social objects in the way that infrastructure is seen as a sociotechnical entity. Another effort is OntoSoft, a distributed software metadata registry that allows researchers to identify, understand, and assess software and then run chosen software and update it as necessary (Gil et al. 2016). Hydrologists find OntoSoft helpful for organizing the metadata associated with software for hydrologic models. However, the process is not seamless and may require the community to be more mindful of crafting code amenable to automated processing (Essawy et al. 2017).

Of course, code availability is only half the battle. For code to be beneficial, it must be possible to compile and run it. Stodden and others (2016) advocate going a step further than code sharing by having journals do reproducibility checks that ascertain that data and other digital objects work. They acknowledge that this would burden journals and reviewers, but it would go a long way toward improving reproducibility. A researcher seeking to replicate another person's code can always read through it and glean some understanding, but compiling and executing error-free code is usually necessary to do other tasks, like running experiments or developing results. Data-processing workflows can be complex and require months to refine, making them nearly impossible to recreate. Much research relies on graduate assistants, postdoctoral researchers, or temporary programmers who may be unavailable for follow-up questions after completing the work. As explored in Section 3.4.2, the primary researchers may not be forthcoming either. It may be difficult to reproduce the operating system, software packages, and underlying libraries in all but the most superficial computing environments.

Methodologies for running code on servers have had mixed success. Projects like Verifiable Computational Results (VCR) are pitched as an approach where anyone can rerun or redo an analysis without dealing with data files, installing or running the software, or dealing with the vagaries of hardware (Gavish & Donoho 2012). This vision for reproducibility centers on developing accessible web services where data and workflows can be permanently stored and accessed. Such systems would host code and plug-ins that researchers could access via standard languages like C++ or R. As with other such efforts, the path forward is unclear. There has been little written recently on VCR, and the associated website appears dormant. Finally, some projects focus on underlying languages and appear to be successful. For example, Cosmes.net focuses on code in well-established modeling frameworks, such as NetLogo, that should be easier to maintain over time and have platform independence because they are implemented with well-known and long-standing Java-based languages.

Another strategy for ensuring the ability to run code is to create self-executing or portable computing packages. Projects including Code, Data, and Environment (CDE) (Guo 2014) or sciunits (That et al. 2017) combine executable code and underlying libraries into a single package that can be moved to another system. Despite their promise, however, these systems are still in their infancy. They are not widely known, let alone used, and it is not clear whether they scale to situations defining big data, such as parallel and distributed workflows. Other efforts center on portable collections of code termed containers and dockers, which are essentially self-contained systems of code that can run software like Python or R with all the necessary dependencies and other code (Klemens 2021). A similar system is CodaLab, which can interpret and run code from various languages while keeping track of the full provenance of an experiment, from ingesting raw data to the final results (Liang & Viegas 2015). The profusion of systems for reproducible science is chaotic but ultimately heartening because it demonstrates the considerable interest in these issues. Section 3.4.4 looks at the growing role of centralized infrastructure in creating and sharing replicable code.

3.4.4 Workflows and Bringing Researchers to Data

Perhaps the biggest roadblock to reproducibility is the fragmented and individualistic research landscape. Moving toward automated workflows and centralized architecture promises to revolutionize data science. Most data and model sharing relies on people spending time and effort adhering to a hodgepodge of dictates and recommendations. The incentives for sharing data and modeling are minimal, and the potential opportunity costs are high. One way to encourage data and modeling sharing is to automate the research workflow from its earliest stages to publication and beyond to develop better research management systems. These would tie into a centralized infrastructure that can provide data, offer persistent identifiers and repositories, and even support later analysis and subsequent publication to support resharing and reproducibility.

There is a deep need for research management systems for even relatively small data science projects. Such a system is usually conceived as a software environment that can record provenance in the form of every data processing step and analytical method applied to data (Peng 2011). These metadata could be part of publications or reside on servers. The Kepler Project was an early entrant into workflow systems designed to allow researchers to develop, run, and share models and analyses (Ludäscher et al. 2006). Kepler draws on local or remote data and integrates existing data via a graphical user interface. Users connect data sources with analytical components to visualize scientific workflows. While other systems have come online, Kepler sees continued use by many projects because it adapts to data science demands, such as integrating new visualization tools (Crawl et al. 2016).

While human-environment data infrastructure tends to originate in environmental work and extend to human domains, there is infrastructure for social data. For example, as discussed earlier regarding administrative data (Section 2.2.3), the Integrated Public Use Microdata Series (IPUMS) projects exemplify data infrastructure. IPUMS has been a leading curator of population and environmental data for the past quarter-century. With over 200,000 users, it is also one of the most extensively used scientific data resources. The project archives censuses and other data, harmonizes them across time and space, develops comprehensive documentation, and makes the data easily accessible (Ruggles 2014). This cyberinfrastructure integrates, preserves, and disseminates data on the human population and environmental attributes (Kugler et al. 2015). These data have a global reach, spanning over 170 countries, global climate data, global land cover and land use, and national and subnational geographic boundaries that allow scholars to attach and integrate other information. Importantly, these global data sets are disseminated to multiple research communities and the public, preserving these resources for future generations. Of course, global data sets do not necessarily imply equal or evenly distributed access worldwide. Accessing and using these free data can require clearing thresholds of expertise and technology (more on these digital divides in Section 5.2).

Many projects see the future of research workflows as being online. Experiment management systems are software stacks and workflows for cloud-based computing that save individual researchers from dealing with the many jobs required to run experiments on research computing systems, such as converting code to run in parallel or distributing jobs across nodes (Monajemi et al. 2019). These systems can be extended to human-environment models and other simulations that can capture, query, and visualize model results (for a review, see Suh & Lee 2018). A prime example of an experimental management system is CyberGIS-Jupyter. This cyber-infrastructure framework uses Jupyter Notebook, an open-source, server-based interactive computing platform standard, to implement geospatial models. Code and data on this platform tie into an underlying application layer that translates the notebook code into code running on the cyberinfrastructure (Yin et al. 2019). Importantly, this framework captures the data and analysis in a workflow into a package reproduced and extended into a cloud-computing environment, which advances the cause of reproducibility. A potential drawback is that these systems may rely on high-performance computing environments. For example, valuable research is done on supercomputing environments at the University of Illinois in the United States. As with other forms of institutional computing, researchers must rely on assurances that the underlying computing infrastructure will be supported indefinitely (Wang 2016).

There is promise in bringing researchers to data instead of the traditional model of downloading data for analysis. Climate simulation is a prime example of where this would work (Networking and Information Technology Research and Development [NITRD] 2016). These simulations are very complicated and increasingly are done on supercomputers or large distributed systems, given how data and computing-intensive they are. For many years, a common practice was to run the simulation on one computing system and then transfer the resulting data to another system for analysis. However, the size and complexity of simulation results make these data transfers challenging to accomplish over the Internet. Researchers can be forced to take unusual steps like copying data onto a *RAID array* (redundant array of independent disks, a set of storage drives that act like one big hard drive) and then physically transporting them to another location.

A better way to do this kind of research is to develop workflows that allow researchers to directly join their models with climate simulations. Alternatively, they could specify in situ analyses where researchers process data in a way that lets them glean, communicate, and store a small extract of the large data set, namely the results and visuals of primary interest. An example is ParaView, an open-source in situ system that extracts data and makes it available via a visualization. This approach allows scientists to explore simulations interactively. They may then choose data to extract for further analysis across various domains, including climate change, landscape change, and seismological modeling (Ross et al. 2019).

One likely endpoint of online research workflows is moving science and scientists toward using centralized infrastructure. Data science should lead the way because most big data work occurs in large, shared environments, including those using cloud computing and high-performance computing. Shared systems allow any university or research lab to focus on offering advances in one area and then share them more broadly (Stewart et al. 2015). For example, Geospatial Data Analysis Building Blocks (GABBS) focuses on developing computing components, including tools, libraries, and data services, supporting user-friendly online interfaces for cyberinfrastructure (Zhao et al. 2017). Geospatial Data Analysis Building Blocks exemplifies the move toward offering reusable software building blocks, in this case for geospatial data management and analysis. It is built on the HUBzero portal platform, which has many natural and biological sciences users. This generic scientific platform allows GABBS to build on an existing computing foundation for storage, management, analysis, and visualization.

A related form of coordination is the creation of science gateways, collections of APIs and web interfaces that allow domain researchers or groups to create entry points into the underlying high-performance computing infrastructure. A science gateway "is usually more than a collection of applications. Gateways often let users store, manage, catalog, and share large data collections or rapidly evolving novel applications they cannot find anywhere else. Training and education are also a significant part of some science gateways" (Wilkins-Diehr et al. 2008, p. 33).

3.5 Focus: Handling Space and Time

There are profound challenges in handling extensive collections of spatiotemporal data. These challenges are germane to big human-environment data and, particularly, the big data of Earth and planetary systems. Despite the fast pace of storage and processing power growth, we face severe bottlenecks in handling many of our existing environmental data volumes. The amount of data is increasing exponentially. Many of the approaches we rely on now for big data, including cloud computing and parallel computing, do not gracefully manage many forms of big spatial data

Spatiotemporal data pose inherent challenges for many reasons. These include complex data models and issues of translating three-dimensional locations to two-dimensional coordinates for analysis. Many standard computing tools, such as parallel workflows that use data indices, are more challenging to create and use for spatial data. Overcoming these challenges involves a range of innovations and strategies that range from trying to accommodate spatiality with existing tools to building spatial capacity into computing frameworks. There is also the development of new fields, such as geospatial artificial intelligence, or GeoAI. This section concludes by examining how the development and use of global-scale land data exemplifies many challenges in handling spatiotemporal data.

3.5.1 Core Challenges

A fundamental challenge in dealing with spatial data is that they are three-dimensional but usually operated on in two dimensions. Spatial data are tied to a location on a bumpy three-dimensional object, the Earth, usually modeled as an ellipsoid (Bolstad & Manson 2022). While some analytical processes use three-dimensional coordinates, most assume locations are on a two-dimensional plane. *Projection* refers to the process and product of moving spatial coordinates from three dimensions to two dimensions. Developing projections has been studied for hundreds of years and involves mathematical functions that map spherical locations to planar ones (Snyder 1987). The tricky part is that it is impossible to do this process without making tradeoffs. Cartographers use the analogy of trying to peel an orange (the Earth) and laying the skin flat (the map). One encounters shearing (stretching the peel in one or more directions), tearing (splitting the peel in places), and compression (bunching up the peel). The result is that no projection can simultaneously preserve a number of basic properties, such as the size or shape of regions.

Projections involve compromises, and these become even more problematic with large data sets. Consider the widespread use of grids to model many environmental phenomena. Much spatial analysis is performed on geometric features such as points, lines, and polygons encoded as vectors or topological graphs (per the data models in Section 2.1). They can be projected with theoretically limitless precision given their zero-, one-, or two-dimensional shapes. In contrast, there are many ways to partition the Earth's surface into a tessellated grid. The square raster cell is the most common, owing to the origin of much data as remotely sensed imagery, but others are possible, usually based on triangles or hexagons.

One challenge with big spatial data is assessing the inaccuracies in global big data, where different tessellations with varying scales can build in subtle errors when calculating distances and areas. These challenges are not insurmountable, but they add another layer of complexity to data analysis. Recent work on discrete global grid systems and other multiscalar reference systems offers a way forward by embracing multiple data models (e.g., raster vs. vector) and assuming that data will come from across scales (Gregory et al. 2008).

Spatial data can be hard to process using parallelization. This workhorse of data science runs into problems with topological or spatial relationships. Many standard spatial analysis methods have not yet been ported to data science or big data. The result is that analyses that should be trivial – such as determining which polygonal areas contain given point locations – are not possible with existing methods or require expensive workarounds (Haynes et al. 2015). Many spatial entities have spatial relationships that define the questions we ask, such as whether a river runs through a given area or ends up in a given body of water. Modeling more sophisticated phenomena, such as cumulative water flow or sedimentation potential at any given point in a landscape, requires tracking watercourse topography.

Maintaining hierarchical, spatial, and temporal associations in a data model or analytical process is challenging, such as tree–stand–forest or person–household–village relationships. Analyzing these entities with multilevel modeling or cyclical graphs requires preserving the entities and their relationships, which can be difficult if shared across nodes. A related challenge is *class imbalance*, when the measurements or objects being acted on are distributed irregularly over computing nodes (Japkowicz & Stephen 2002). Spatial data often pose this challenge because there is significant variation in the size of mapping units being processed. Some nodes deal with large data chunks while others are idle. This issue affects many newer data science approaches (Buda et al. 2018).

There is exciting work on overcoming specific problems posed by big spatial and temporal data. Researchers working with remotely sensed imagery are developing new ways to do essential tasks such as mosaicking images to form single data sets. They also use well-established big data platforms like MapReduce and Hadoop to implement foundational spatial operations (Li et al. 2019). Even reasonably straightforward methods, like statistical regression or Monte Carlo analysis, are the object of ongoing attention as they are implemented in parallel processing environments (Crosas et al. 2015). Measures such as ensuring that data and processing jobs are matched to underlying computing node capacities are one of many ways to squeeze out additional performance (di Modica & Tomarchio 2022).

Visualizing big data is also an open area of research in general, and it is vital for spatiotemporal data. Spatial science can draw on a long and rich history of geovisualization, starting with cartography and broadening in recent years into exploratory data analysis and complex dynamic visualization. However, most of this work is performed on data sets orders of magnitudes smaller than those that characterize big data. Big data are often two, three, or four dimensions, and many traditional geographic information systems are firmly focused on two dimensions with few exceptions (Li 2020b).

Spatial data science involves a good deal of fundamental work on improving spatial indices and functions. Many forms of spatial analysis rely on a spatial index to keep track of spatial entities and how they are related to each other. Development and implementation of spatial indices have long been a research focus in spatial information science and are undergoing a renaissance with big data (Li et al. 2017). Data science also requires simple spatial functions that can work with indices with topology, such as determining whether two lines connect or whether a point lies within a polygon, or spatial adjacency, like taking the mode of a group of raster cells (Ray et al. 2015). More advanced functions can then be derived from these simpler ones, such as spanning a river network or establishing a *viewshed*, the parts of a landscape visible from a given vantage point. Indices and simple functions are integral to queries, ranging from finding spatial entities that meet some simple criterion, like size or attribute, or more complicated forms of data mining (Yang et al. 2017b)

Spatial data implementation approaches can be on-top, from-scratch, and built-in (Eldawy & Mokbel 2016). On-top approaches are the most straightforward and usually least efficient because they draw on an existing system that deals with nonspatial data. Spatial queries are built into a separate layer of programming that has to rely on workarounds to deal with spatiality. On-top is the most common approach to spatial big data and figured prominently in early work in developing core spatial functions such as spatial indexing, tessellations, and creating k-mean metrics (Prasad et al. 2017).

These systems treat spatial data like any other data or rely on standard models like matrices or graphs to represent raster and vector data. Using standard data models is adequate for simple analyses, but many forms of spatial data analysis rely on specialized data models. Similarly, using existing nonspatial big data systems means that new spatial analytical functionality must be built on top of existing projects. Fundamental vector processes like the polygon unions, Thiessen/Voronoi polygons, and convex hulls are relatively recent additions to the MapReduce family in Hadoop (Li et al. 2019).

From-scratch and built-in systems are sometimes entwined. From-scratch approaches involve building spatiality into a high-performance computing system from the ground up. This method is fast, efficient, and potentially hard to maintain (Wang 2010). A few from-scratch systems are in everyday use, including SciDB and Rasdaman, both designed for scientific applications requiring multidimensional arrays suitable for raster analysis or handling massive image data sets. One such system is EarthDB, which uses SciDB to handle vast collections of satellite imagery (Planthaber et al. 2012).

Researchers often use the built-in approach to strike a balance between on-top and from-scratch methods. As the name suggests, built-in systems carefully weave spatial data and operations into existing systems. Several exciting projects build spatial implementations into established big data ecosystems, such as SpatialHadoop, Esri GIS Tools for Hadoop, and Hadoop Geographic Information System (Eldawy et al. 2016). There is also interest in extending this work to new models such as Spark's resilient distributed data set platforms that offer more speed and fault tolerance (more on distributed and parallel computing in Section 3.3.3).

The state of spatial computing is in flux. A mix of extending existing methods is deployed alongside bespoke approaches against a backdrop of fast-paced changes in the underlying methods (Shook et al. 2016). Storied warhorses like Hadoop serve as the foundation for methods like SpatialHadoop or new systems that can fully leverage standards like Hadoop. More recent entrants, such as Spark, offer developers trade-offs. Is it better to use well-established methods, like PostGIS Raster, on an established but potentially limited system? Take a gamble on a new system like Spark that promises much more capacity, scalability, and speed at the cost of working in a new and comparatively shallow ecosystem (Hu et al. 2018)? The open-source movement offers hope because it focuses on both new advances and developing common standards that allow researchers to ignore the underlying technology (to some extent) and just focus on interfaces and standards. The OGC offers a family of standards, including the Web Feature Service (WFS) for serving vector data; Web Coverage Service (WCS) for raster data, point clouds, and meshes; and the Web Map Service (WMS) for visualizing data quickly on multiple devices (Baumann et al. 2016).

There is growing work on spatially explicit artificial intelligence, often termed GeoAI. *GeoAI* is short for geospatial artificial intelligence, a blend of artificial intelligence, spatial big data, and high-performance computing (Li 2020a). GeoAI builds spatial knowledge into tasks such as ontology creation, using spatial rules to help develop semantics. For example, spatial facts like rivers flow to oceans or straits connect large water bodies can be used to help distinguish among similar geographic names or enable complex spatial queries that rely on topology. GeoAI is increasingly central to human-environment scholarship, as seen in applications such as deep learning to generate fine-scaled land cover maps from coarse spatial resolution images (Ling & Foody 2019). GeoAI is also finding its way into substantive domains like environmental epidemiology, including modeling air pollution or infectious disease spread (VoPham et al. 2018).

3.5.2 *Global-Scale Land Use and Land Cover Data*

Global-scale land data are essential to human-environment research but exemplify many challenges in handling spatiotemporal data. These data sets are central to much progress in human-environment research, including land change science, global environmental change, hazards, and ecology. Data science approaches are increasingly used to develop new data sets, including big data workflows, cloud computing, machine learning, and other methods. These approaches are also used to integrate existing data sets and handle challenges, including developing onto-logical classifications for land use and land cover.

Land data are central to a broad swath of human-environment scholarship. Exemplars of global land data sets include Global Land Cover, developed as part of the Millennium Ecosystem Assessment by the United Nations (UN). This product draws on existing land cover databases, remotely sensed imagery collec-tions (especially SPOT 1-km vegetation sensor), and expert knowledge about ground conditions (Bartholome & Belward 2005). Another is the IGBP DISCover global land cover classification developed under the aegis of the International Geosphere-Biosphere Programme. It uses clustering and postproces-sing on normalized difference vegetation index (NDVI) values from Advanced Very High Resolution Radiometer (AVHRR) data (Loveland et al. 2002). More recent data sets include Global Land Cover SHARE (UN Food and Agriculture Organization [FAO]), Global Land Cover by National Mapping Organizations (GLCNMO), GlobCover (European Space Agencies and others), GlobeLand30 (National Geomatics Center of China), LC-CCI (Land Cover Climate Change Initiative), and MODISLC (NASA, Boston University, and others) (Laso Bayas et al. 2017). These land use and land cover data are essential to answering many pressing questions about human-environment dynamics.

Data science approaches are central to developing new global data sets. Remote sensing remains integral to this work. Remotely sensed imagery is incredibly useful given its spatial and temporal qualities, but also because, as noted in Section 2.4, remote sensing overlaps with data science. This coevolution has led to data science workflows and methods in remote sensing while also bringing scientifically essen-tial issues and projects to data science (Wulder et al. 2018). Global data sets are massive because data with even modest spatial or temporal resolution is measured in terabytes, and they can be represented via different data models.

Global data inherit all of the methodological challenges posed by spatiotemporal data. For example, GlobeLand30 data offer good attribute resolution, 30 meter spatial resolution, and 80 percent accuracy via a data science workflow approach (Chen et al. 2015). Another product, the Global Urban Footprint, uses machine learning on vast holdings of over 180,000 TanDEM-X and TerraSAR-X radar images that offer 3 meter ground resolution (Esch et al. 2017). These maps, primarily by virtue of their resolution, offer an average accuracy of 85 percent, comparable to the best existing data sets. Cloud computing also promises researchers a way to process vast amounts of existing and incoming data. For example, Google Earth Engine was used to process images from one of the largest and longest-running collections of remotely sensed imagery, the Landsat Thematic Mapper archive, to create fine-scaled classification using fairly standard supervised and unsupervised classification (Azzari & Lobell 2017).

Developing global land data sets increasingly takes a data science approach to integrating multiple data sources. Creating land data sets often involves incorporating independently derived data sources such as high-resolution satellite data, aerial photography, and economic and census-based surveys at various aggregated scales and resolutions. Machine learning is a valuable tool for combining and refining data sets and an active area of research because the messiness of the real-world phenomena measured by remotely sensed imagery is challenging to many approaches (Maxwell et al. 2018). Base data are used to develop derivative products such as texture, pattern, and degree of spatial dependence. These can be used to estimate features such as population or settlement (Leyk et al. 2019). Ancillary spatial information can be helpful as secondary predictors and accentuate relationships not detectable or initially obvious with remotely sensed pixel observations. Of course, machine learning is not a panacea for this kind of mapping. Garbage-in-garbage-out still applies, and it may be that many global land cover products are less accurate than presumed because they draw on a range of data sources that are often biased in temporal or spatial coverage (Meyer & Pebesma 2022).

Crowdsourcing and big data workflows help develop land use or land cover observations. The Geo-Wiki, for example, is managed by researchers with the International Institute for Applied Systems Analysis (IIASA) in Austria. This citizen science platform offers thousands of volunteers a way to provide feedback on existing information or develop new data via a computer or mobile device (Laso Bayas et al. 2017). Geo-Wiki data acts as a land cover and land use reference data set that may be used to calibrate and validate global land cover. For example, Geo-Wiki was integral to developing a global map of forest extent. This project combined forest statistics from the UN FAO with global land cover data and crowdsourced information to develop a more accurate data set (Schepaschenko et al. 2015). As with other forms of big data, the validity and utility of these products are enhanced when tied to field-based measures, expert knowledge, and biophysical models that capture important features of trees and tree stands (Réjou-Méchain et al. 2019).

Ontological issues abound with land data. There is a lot of time and effort dedicated to defining, measuring, and classifying seemingly simple land cover categories like forest or grassland. Developing a global land cover data set is ultimately an ontological challenge that quickly transitions into issues of algorithm and measurement. What is a forest? What is a grassland? Each specific data set embeds definitions that are a mix of expert judgment and a translation of measures into discrete categories or as measures of membership, such as the percentage of a given cover class. There is a good deal of interest in developing new ways to use machine learning and other core data science approaches to create global-scale land data sets from sparse and robust training data sets (Robinson et al. 2021). Remote sensing has long dealt with extracting simple categories, such as forest or bare soil, from complex and noisy remote sensing data (Ahlqvist 2008). This work tends to represent category semantics using formal axioms including hierarchies of categories, such as different kinds of tree cover within a broader category of tree cover. They also draw on prototypes or exemplars where the system classifies objects based on their characteristic properties (more on ontologies in Section 3.3.2).

Data science approaches are increasingly used for data set fusion or reconciling differences among data sets to create hybrid coverages. Fusion or hybridization is challenging because it entails overcoming foundational differences in sources, processing, and purpose (Fritz & Lee 2005). Much research flows into combining and comparing different data sets. As with many other kinds of data, most efforts conclude that fitness for use can only be determined by examining how the data were developed and their eventual use. For example, testing the consistency of five commonly used land cover data sets (GLC2000, CCI LC, MCD12, GLOBCOVER, and GLCNMO) finds the overall consistency among all five data sets ranges was middling, from about 49 percent to 68 percent. However, some regions fare better than others, typically those with better underlying data like Europe. Machine learning can fuse these disparate kinds of global land cover data sets, drawing on high-resolution regional data sets, to create a new and more accurate data set (Song et al. 2017).

3.6 Conclusion

Data science offers new ways to engage with big data of human-environment systems. Spatial data are valuable for understanding nature–society dynamics, but at the same time, these data pose unique challenges to their use. Existing and future research directions specific to large spatiotemporal data sets hold much promise for data science because they spur new developments. These include the move toward resilient distributed data sets that can be more readily analyzed and the creation of specialized hardware and software. These advances help lift the roadblocks in working with complex data in parallel, cloud, and distributed computing. There are also exciting new approaches such as advanced computer-aided visualization tied to machine learning algorithms that help researchers make sense of the vast new data sets.

Engineering, computer science, and cognate fields are racing to develop solutions to problems along all points of the data pipeline. These include the design of new sensors that expand the kinds of measurements taken of many phenomena, such as satellite-based sensors for greenhouse gases or inexpensive ground-based monitors for pollutants and trace gases that have environmental and human impacts. There is also the ongoing development of new, automated ways to clean, collate, and otherwise simplify data in situ before it is stored upstream. These approaches promise to reduce the volume of big data while also improving the data we possess. Also essential is linking the data science of human-environment systems to other fields, including better integrating research on data lifecycles and associated trappings around metadata, provenance, and ontologies. These data features make much of the core computational work in data and advance the cause of reproducible science.

4

Theory and the Perils of Black Box Science

Data science proponents sometimes contend that their approach augurs the "end of theory" because we are on the threshold of a new scientific paradigm. Big data is seen as a powerful "black box" into which users shovel large and messy data collections and, in return, get deep insights into any domain people care to investigate. The essence of black box research is that the inner workings of the computation can remain opaque to most users, and there may be little need to understand much about the process being modeled. Data science can pursue an inductive approach to knowledge creation that frees researchers from fieldwork or dealing with the messy business of deductive science that involves developing hypotheses and conceptual frameworks. The proposition of theory-free science has sparked great debate and attendant conversations on how science is practiced and moving beyond simple epistemological binaries.

There is an emerging middle ground in the debates over theory development in data science. Researchers of all stripes increasingly combine data science with other analytical modes to advance knowledge. There is growing recognition of the value of domain knowledge and small-data studies in data science and how data science can offer value to other research domains. There is also much discussion about the skills and content of data science. Many of the issues around theory development in data science and the perils of black box science play out in the research and practice of smart cities. Computational approaches are central to the vision of building theory into a science of cities that can also have positive real-world impacts on people and the environment, although caveats abound!.

4.1 What Is the Science of Data for Human-Environment Systems?

Science is often treated as a monolithic entity when it is a multifaceted and protean enterprise. Discussions of the pros and cons of data science for human-environment research revolve around the tension between inductive and deductive modes of science. However, scholarship in practice is messier than these positions imply. Indeed, the proposition of a theory-free science is profoundly at odds with the core conceptual precepts and scientific practices of many disciplines that seek to advance the understanding of human-environment systems. However, science is done by different people in different ways. It is helpful to position data science and other forms of scholarly inquiry in this larger arena. There is also value in piecing out how different knowledge domains think about causality, prediction, and explanation in the context of data science. The capacity of modeling to support theory development is contingent on fundamental issues around sampling, bias, data, and those having to with model form and assumptions. Doing science well in the data science era is often more complicated than it appears.

4.1.1 Reframing the Epistemology of Data Science

Data science for human-environment research must be seen as part of a larger conversation around how to do science. As explored earlier, data science is essentially about developing models from data for descriptive, explanatory, or prescriptive purposes. Big data has added tools and complexity to this drive, but for centuries, the underlying focus has been to develop models from data. Epistemological framings around data science are often focused on the tension between idealized notions of inductive and deductive work. This dualism oversimplifies scholarly inquiry, but it does offer a lens onto some key issues.

For many scholars, the idealized version of science is the deductive model, where one formulates a hypothesis in a form that can be falsified by testing it against data. This model (also termed hypothetico-deductive) is simplified because most science is messier in practice, allowing for continency, underdetermination, and indeterminacy. As explored further in this chapter, there are other valid ways of knowing. However, the deductive approach is a useful shorthand for the general practice of science that contrasts with inductive science, where the researcher moves from data to developing models.

Data science, for many researchers, comes with a primarily inductive epistemology that is at odds with the underlying theory-seeking orientation of much scientific work. A much-publicized article in the popular technology magazine *Wired* argued that big data is a harbinger of the "end of theory" (Anderson 2008). *Wired* has long been happy to pronounce and prognosticate for its readership, seeing as it is interested in the intersection of society and technology. Given the venue, the author could be forgiven for engaging in hyperbole in assessing what big data means for theory. Notably, he is not writing in a vacuum because many people share the contention that big data offers a powerful and flexible black box approach for creating knowledge.

The corollary to the end of theory is that it is increasingly possible to generate new knowledge in many fields – urban systems, climate change, ecology – without drawing on domain expertise or engagement with existing research. Indeed, such an approach is considered by some to be desirable because it frees researchers to explore a topic unbound by previous research and unblinkered by domain knowledge and norms (Berry 2011). Data science proponents contend that big data offers resolution and depth previously unseen in most fields and insights or rules without human bias or foibles.

The debate over inductive versus deductive methods is the most recent phase of a discussion that has been occurring for centuries. While data science is often touted as paradigm-breaking or a fundamental shift in how humans understand the world, it is important to situate these claims in the broader history of science. "Modern science was born in the seventeenth century as a fusion of observation and reason. Radical empiricism (data without reason) and rationalism (reason without data) were rejected in the quest for knowledge of Nature" (Coveney et al. 2016, p. 2).

There are many way stations on the journey between the poles of raw empiricism and pure rationalism. The modern research enterprise is pluralist in its approach, even when drawing on scientific practices that trace back centuries. While sciences in the abstract may teach that theory-building is straightforward (e.g., starting with empirical observations, deriving testable hypothesis, and then iterating), many scholars do not necessarily seek to start from or build grand or totalizing theories, but instead bound them by space, time, or social construct while also seeking to bridge across different explanations (Boudon 1991).

Some scholars argue that data science critics are missing the point. It is unproductive to focus on failures in not adhering to scientific precepts, including theory development. Instead, engineering better and more efficient predictive models can occur alongside other methodologies. "Big data are no substitute for hypothesis-driven scientific discovery, and many of the types of data gathered are dependent on the idiosyncrasies of the data collection system. However, I contend that these criticisms are largely irrelevant when the objective is to build an effective computational artifact" (Lin 2015, p. 39). In this view, data science methods are very good at identifying interesting connections in data, finding subtle patterns, or answering specific questions. Sometimes these are goals in and of themselves, and other times they are pursued in service of advancing knowledge, in which case they would be employed by domain scientists or by data scientists in collaboration with others. The essence of machine learning, for example, is prediction under stationary conditions (Section 4.1.4 dives into issues around model form and assumptions) and not causality (Athey & Imbens 2019).

A realistic model of science is an iterative one that incorporates a broad array of different and occasionally seemingly antagonistic approaches. Here the goal is not some platonic ideal of hypothesis testing nor blindly opportunistic data trawling, but instead thinking about approaches in terms of how they contribute to overarching research programs (Elliott et al. 2016). The nature of the knowledge and motivation involved will interact with the research framing, which can range from hypothesis testing through more general questions or expectations to achieving a specific goal (Figure 4.1). This framing interacts with a range of methods,

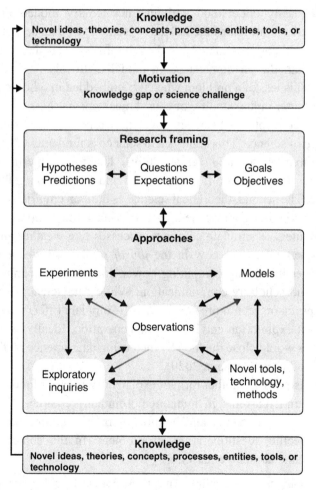

Figure 4.1 Forms of scientific inquiry (adapted from Elliott et al. 2016). Reprinted with permission from Oxford University Press.

including observation, modeling, experimentation, explanatory inquiries, or methods that have yet to be developed. Evaluating the quality of work is on

the basis of the alignment between the nature of the knowledge gap or challenge addressed and the combination of approaches or methods used to address the gap. Research should be evaluated favorably when it incorporates approaches and methods that are well-suited for addressing an important gap in current knowledge, even if they do not focus solely or primarily on hypothesis testing.   *(Elliott et al. 2016, p. 883)*

An expansive view of science allows for competing schools of thought on how we know what we know. A leading candidate for an underlying epistemology of data-centric and model-intensive inquiry is the *semantic conception* of science (Henrickson & McKelvey 2002; Portides 2017). Under this approach, the data and theory of many actual human and environmental systems are too opaque or complex to be easily understood without intermediary models. The semantic conception of science mitigates the inductive leanings of data science by positing that we develop a theory by linking big data to other models that bound, context-ualize, and explain the phenomena being modeled. In this view, the black box nature of a model is less of a problem when it is embedded in other models that can help develop a better understanding of causal mechanisms (Manson 2007).

The semantic conception of science dovetails with what data scientists call the fourth paradigm of science. This approach to science is fundamentally distinct from the previous three phases (Hey et al. 2009). The first, experimental science, flourished in the period before the renaissance and focused on gathering data to describe nature. The second, theoretical science, is built on empiricism to move into law-seeking via modeling and generalization. The third, computational science, leveraged computers to simulate complex processes like weather and climate. In this view of science, we are now in the *fourth paradigm*, where massive data volumes and ever-increasing computing power dramatically change the ways of doing science and scholarly communication. While data science may be seen as a return to empiricism, seen more constructively, big data offers new and exciting methods for data exploration and hypothesis generation. Ideally, when used well, these approaches would close the loop by then using data science methods to prove or disprove hypotheses (Gahegan 2020).

Abductive reasoning is another form of explanatory framing for data science that captures many undercurrents in human-environment research. These forms of analysis focus on the dynamic and helpful interplay among data, method, and theory for a specific question, issue, or problem. In the context of complex spatiotemporal data, abductive reasoning

requires four capabilities: (1) the ability to posit new fragments of theory; (2) a massive set of knowledge to draw from, ranging from common sense to domain expertise; (3) a means of searching through this knowledge collection for connections between data patterns and possible explanations; and (4) complex problem-solving strategies such as analogy, approximation, and guesses. Humans have proven to be more successful than machines in performing these complex tasks, suggesting that data-driven knowledge-discovery should try to leverage these human capabilities through methods such as geovisualization rather than try to automate the discovery process. *(Miller & Goodchild 2015, p. 457)*

The primary advantage of abductive reasoning is that domain experts can monitor and engage with big data to dismiss obvious, ridiculous, or meaningless models. Additionally, it allows these same domain experts to engage with data science instead of dismissing it.

Simulation modeling is central to abductive and semantic approaches to knowledge. Models capture some of the benefits of big data without losing sight of understanding the processes involved in the modeled systems (O'Sullivan 2018). In this view, the critical constraint of using big data is an oversimplifying focus on establishing correlation and classification over developing models of causation. This oversimplification is exacerbated by using temporal snapshots collected, often for unrelated purposes, as a stand-in for processual understanding. In contrast, theory-led explanations focus on models that support open-ended processual explorations.

Simulation modeling also allows for a greater range of contingency and inherent uncertainty about the nature of knowledge. It embraces the need to bring to bear multiple methods, approaches, and perspectives to understanding many (if not most) human-environment systems (Manson & O'Sullivan 2006). Linking process-oriented or mechanistic modeling to the inductive or statistical models central to big data can lead to promising results. This coupling of big data to various forms of modeling addresses the critique that big data can provide the "what" but not the "why" or "how." Section 4.3 examines how data science approaches play out for specific domains and dives into some of the specifics of these linkages in several specific research areas.

4.1.2 Prediction, Explanation, and Causality

Entwined with epistemological issues are questions of why we model the world. Data science and scholarly inquiry, in general, can be considered in terms of the general rationale for modeling. Is the model descriptive, normative, or prescriptive? Is the model trying to predict or explain or determine causality? Much has been written on the nature of explanation – centuries of careful consideration – so this section does not attempt to capture that long conversation. However, one helpful framing of the challenge of doing data science well is that any kind of science involves tension between prediction and explanation, and how this tension plays out in terms of causality.

Many disagreements around data science center on the tension between prediction and explanation. *Prediction* is about developing effective methods to ascertain the future states of a modeled or real-world system. *Explanation* is about developing approaches that lend insight into the relationship among variables that describe a modeled or real-world system or capture some understanding of processes that the variables represent. Figure 4.2 offers a two-by-two matrix of prediction versus explanation that distinguishes among description, prediction, theoretical explication, and exegetical construction. There are other ways of thinking about explanation versus prediction but these four capture essential elements.

Consider these four categories in the context of data science.

- Descriptive exposition is low in predictive and explanatory power but is the foundation for an enormous amount of knowledge in human-environment systems. Being able to describe a system reliably requires a lot of data and exploratory data analysis. Descriptive modeling describes a system without necessarily claiming that this portrayal is valid for all times or places.
- Predictive analytics do not necessarily seek to build coherent theoretical explanations, instead focusing on making good predictions. The classic black box concept applies to predictive analytics because knowing what is happening within is unnecessary as long as the box works. Commercial entities are particularly invested in the predictive aspects of data science. A model that improves predictions of input costs or consumer demand by 1 or 2 percent can translate into substantial money.

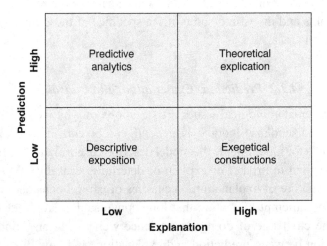

Figure 4.2 Prediction versus explanation (adapted from Waller & Fawcett 2013). Reprinted with permission from John Wiley & Sons.

- Exegetical construction takes a page from the humanities by providing a coherent reading of a situation, often a qualitative-oriented or qualitative–quantitative model that moves from description to a descriptive (but not necessarily predictive) explanation. The term model is loose in that it could mean a lengthy essay with some numbers or a complex agent-based model that encodes variables and relationships.
- Theoretical explication implies that an explanation of a phenomenon is so apt that a model based on it will accurately predict future system states. This approach is typically seen in the natural sciences, where theory explains and predicts real-world phenomena. This category of explanation tends to be rare in the ecological and social sciences.

Theory development is usually done by establishing explanations that support claims of causality, although prediction has its place as well. Work that is more predictive than explanatory can open the door to theory development when it spurs the search for why a given predictive model seems to perform well. While correlation is not causation, tracking correlations among variables of interest can undoubtedly point toward new and interesting relationships deserving of study. Explanatory modeling is a key part of developing causal explanations. Most approaches to causality rely on the assumption that cause precedes effect. Two simultaneous events may have the same cause, or they may be related when one event routinely follows another. Translating these simple ideas into statistically and epistemologically defensible models is challenging. The way forward for much of science involves a close and more nuanced examination of a model "by adding more realism and being more careful who asks what question, using what information, and when the question is asked" (Granger 1988, p. 551).

Establishing causation in human-environment systems is a fraught yet long-running enterprise. Many core human-environment processes and patterns are only indirectly captured by many of our standard observation methods. For example, remote sensing is an essential human-environment data source (Section 2.4). However, it is difficult to move from raw spectral signatures to land cover categories to land use classes to socioeconomic or environmental considerations (Crews & Walsh 2009). This need to develop chains of analysis and relationships has forced researchers to carefully assess commonly used sources of information such as remotely sensed information or census data while also teasing out new ways to understand coupled human-environment systems. Environmental sciences wrestle with similar issues. The rise of big data in macroecology, for example, has been accompanied by a renewed focus on developing theories and mechanistic understandings coupled with inductive data science approaches to understand better how species change over time (Wüest et al. 2020).

Advances in modeling causality are spread among fields. There is a common interest in mathematics and engineering to use a mix of engineering and statistical methods to firm up approaches to causal approaches. Much recent work focuses on using graph-based models that can create complex models of probability, including Bayesian networks of conditional probabilities (Pearl & Mackenzie 2018) and extensions to standard structural equation models in the social sciences (Morgan & Winship 2014) and in bioinformatics (Meinshausen et al. 2016). If the causal structure is thought to be known, then the model can be used for prediction, and if the structural relationships are unknown, the focus is on discovering them. These methods can merge into more complicated processual models, like simulations, but these often boil down to structural relationships and variables. The philosophy of science and studies of science and technology (Section 1.5) are also deeply invested in discussions around the role of data science in modeling causality.

Data science applied to coupled human-environment systems opens new avenues for understanding and establishing causality. The overarching goal of this work is to assess whether they may be causally related to each other or share a common underlying driving variable. Finding these relationships involves testing for unobserved variables and dealing with sampling problems, measurement error, nonstationarity, and nonlinearity (Runge 2018). Related approaches focus on opening up the black box of machine learning and other data science methods to make the processes less opaque.

These methods may be combined with deep learning or other big data approaches to further scientific understanding of Earth systems. For example, geosciences have pushed data science forward by finding success with soil mapping from sparse data, examining carbon dioxide fluxes, and modeling terrestrial photosynthesis (Reichstein et al. 2019). The ultimate success of data science work in human-environment systems will likely involve linking patterns to processes. O'Sullivan (2018) argues that when people focus on the velocity of data, they often incorrectly conflate the rush of a series of snapshots with a deeper understanding of underlying processes. Many data science methods focus on relatively static conceptions of the world because they examine slices in time. These methods should be used in conjunction with a conceptual view of the processes to allow exploration of dynamics and processes.

4.1.3 Poor Data, Bias, and Sampling

Poor data often bedevil data science. The most fundamental challenge of data-centric science is that often the data are not sufficient for many forms of modeling. Data science is no different from other approaches in often needing more, or better, data than anticipated. It is increasingly evident that big data approaches often need vast amounts of good-quality data to develop defensible statistical models. Even then, these models may be wrong. Even a sophisticated data science model can be overfitted or under-specified as with other analytical approaches, especially in how these models often ignore underlying processual or structural characteristics of the real-world system. Issues arising from data quality can be exacerbated by bias and misapplying concepts around sampling. At the same time, data science is proving helpful in making human-environment data better by illuminating problems and filling gaps.

Many of the pitfalls in big data lie in underemphasizing essential statistical concepts. Statisticians are increasingly militant in pointing out that statistical approaches drive much data science and that these methods are sometimes poorly executed (Donoho 2017). This shortcoming is partly due to too much emphasis on data modeling and handling; while an important topic, it is only one of many in data science (Section 1.3). Consider the issues of sampling vis-à-vis analysis. Data sampling is central to statistical analysis and many human-environment system domains. For most problems, whether understanding climate impacts on commu-nities, the nature of tree–stand dynamics, or how shellfish respond to shifting currents, researchers almost necessarily rely on sampling, given the time and cost required to investigate these problems. There is a deep body of work on power analysis, for example, to determine how large a sample must be to support a given analytical approach.

Data science often looks to avoid sampling issues. Assuming that all of the data on a population are available obviates the ostensible need for a sample. As discussed in Section 2.3, big data have systemic biases, shortfalls, and blind spots like any other kind of data. At issue is the assumption that, as big data get better, the sample size approaches the size of the population in a classical statistical sense. However, this is rarely true. Sources are often bounded in space and time. No data are unbiased in simple measurement terms; even sophisticated remote sensors only collect a range of spectra. Most ground-based environment monitoring stations collect a fraction of the measurements possible. Other data have more insidious forms of bias, such as socioeconomic bias in social media, or governmental or commercial biases in remotely sensed imagery. These biases ensure that these data do not map onto all populations. People, places, or other observation units are misrepresented or not included. Calculating the nature of this bias and how it maps onto even simple measures of central tendency like the "average" person in the data set is a fraught enterprise because big data are never actually exhaustive or fully representative of a population (Meng 2018).

Data sets are often repurposed in ways that impair their fitness for use. Even if research samples are so vast and representative as to allow sampling issues to be ignored, most are usually collected for one purpose or on one platform and then used to examine entirely unconnected phenomena. Researchers often reassemble data collected by someone else to meet goals that may differ from or even conflict with the original researcher's (Ruppert 2013). Repurposing data is tricky in these cases because the sample is not intended for its new use. Even the seemingly straightforward task of converting millions of simple temperature measurements to meaningful descriptions of average global surface temperature is a vastly complicated undertaking. These data vary from place to place and over time and draw on many different sources (Rohde & Hausfather 2020). Automating these data-to-information transformations through machine learning and data-mining tools is an open research area.

Spatiotemporal human-environment data exemplify big data's "variety" criterion. They are often limited in their temporal extent and have a good deal of noise. For example, data science applied to climate change confronts messy data from a broad array of sources, including paleoclimatic records, model outputs, and modern-day climate and weather measures (Faghmous & Kumar 2014b). Some of these data are suited to the vacuum cleaner nature of big data. Data sets for precipitation or temperature are relatively well-formatted, reasonably complete, and have a simple data model. Like the paleoclimatic record derived from multiple approaches, others require a deft touch and domain knowledge to make sense of them. Preprocessing and manipulation are necessary before working with even the most basic data sets. Data scientists must choose between the data-handling challenges of raw data and the ease of clean data that have been manipulated in ways that can introduce bias or hidden correlations that result purely from the processing (e.g., using elevation to correct for error in temperatures can introduce multicollinearity).

Data science approaches help tackle uncertainty in human-environment data. For example, our capacity to develop land cover data from satellite remote sensing increasingly draws on machine learning that uses ancillary data such as slope, elevation, or crowdsourced information (Section 2.2). The use of cloud-based systems and other advanced forms of computation has also spurred growth in analysis that draws on an entire library of images, not just dozens of carefully chosen snapshots typical of this work (Wulder et al. 2018). One such example employs machine learning techniques to accommodate data with different scales to help correct errors in remotely sensed imagery gathered by many low-cost satellites (Houborg & McCabe 2018). Section 2.4 delves into other examples of data science that is being used to combine data sources to combat uncertainty.

In the extreme, it may be necessary to embrace the idea that some forms of uncertainty can be countered while others cannot. Couclelis (2003) outlines and distills various forms of error and uncertainty in data:

that which is unknown or unsuspected; that which is not understood; that which is undiscussed or undisputed; that which is deliberately hidden; and that which is distorted. To these we may add at least one more: that which one does not want to know . . . But what about that which cannot be known? What about uncertainty that is not a matter of incomplete, distorted, misunderstood, or mishandled information? *(p. 167)*

An essential contribution of this work is distinguishing between information and knowledge and establishing that many forms of uncertainty cannot be corrected.

Some questions are unanswerable, and some phenomena are unknowable. The average data science workflow is unlikely to encounter the challenges found in the more abstract realms of the philosophy of knowledge, be it Wittgenstein's limits on the use of language, Gödel's incompleteness theorems on the limits of any axiomatic system, or Heisenberg's uncertainty principle. Nonetheless, these and other conundrums undergird challenges of uncomputability and undecidability in computer science, and challenges in capturing the spatiotemporal and attribute characteristics of most real-world phenomena. To these are added the various social and political challenges to collecting and sharing data examined throughout this volume. Recognizing the myriad forms of uncertainty is not meant to be the counsel of despair nor an invitation to throw our collective hands up in the face of intractable error and uncertainty. Instead, scholars and other data users must embrace error and uncertainty and find ways to do data science on coupled human-environment systems.

4.1.4 Model Form and Assumptions

Most data science methods can be understood in functional terms. Some outcomes are treated as a function of input data, where the function may be some sort of correlation, a means of classification or grouping, or pattern matching (Section 3.1 examines core data science tasks). More generically, many implementations of this formulation borrow heavily from standard statistical concepts, such as linearity or correlation but highly dimensional and complicated human-environment data often challenge these assumptions (L'Heureux et al. 2017). These data have spatial and temporal characteristics that can violate many underlying statistical and modeling assumptions adapted for some kinds of data science (per Section 2.3 on bias and error).

Human-environment data are multidimensional in ways that lead to spurious correlations. There are so many variables across so many data points that any naïve modeling method can find seemingly significant relationships. At its simplest, there is a nontrivial chance that two variables are correlated due to chance in an extensive collection of variables. Such correlations can be statistically significant but scientifically irrelevant or lead to biased big data models (Fan et al. 2014). Caldwell and others (2014) performed one of the earliest data science analyses for climate data. They demonstrated how machine learning approaches could produce seemingly significant but ultimately meaningless results that ignore statistical issues. The authors used machine learning to automate a search for relationships across outputs for an ensemble of climate change models to find predictors for changes in climate in response to changes in underlying phenomena such as atmospheric carbon dioxide. They found that such analyses are inadequate when they do not account for dependence among differing model configurations and variables or the effect of locational differences or seasonal variations. Data mining yielded tens of thousands of seemingly significant relationships, but all were explainable by chance. The authors concluded that data mining approaches are helpful primarily in terms of identifying potential relationships. These relationships can in turn feed into carefully designed investigations of actual climate systems and mechanisms that are grounded in physical measurements and experiments.

Endogeneity can be found in many human-environment data sets and confound analyses. In simple terms, *endogeneity* occurs when the residuals of a model correlate with some predictors and can lead to misspecification. In a regression model, for example, endogeneity leads to biased estimates. More broadly, endogeneity occurs when a bidirectional relationship exists between output and input variables. Land change analysis, for example, confronts a classic issue in modeling the relationship between deforestation and road development, since they have a circular relationship in that more logging leads to more roads, which can lead to more logging (Busch & Ferretti-Gallon 2017). The existence of highly dimensional spaces in big data makes it more likely that there will be incidental endogeneity, where one or more predictors will map onto residuals because there are just so many predictors in play (Krishnan et al. 2019). Statistics and cognate fields like operations research have been at the forefront of expanding existing methods or developing new ones to deal with these issues in a data science context (Fan et al. 2020).

Autocorrelation is a significant source of issues and opportunities in many data science operations. *Autocorrelation* occurs when observations related over time or space are not independently distributed but instead conditioned on neighboring values. Methods relying on this assumption can produce incorrect measures of significance. For example, climate data are stored and analyzed as grids or matrices that capture the inherent spatial and temporal autocorrelation of many natural

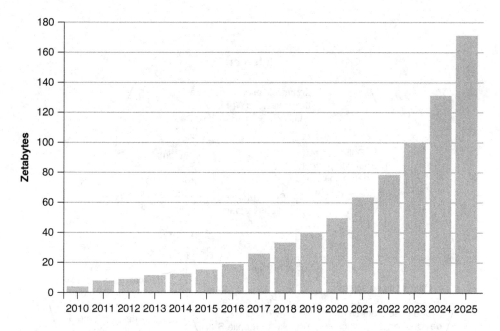

Figure 1.1 Size of global data holdings (2010–25) (Reinsel et al. 2018). Reprinted with permission from the International Data Corporation.

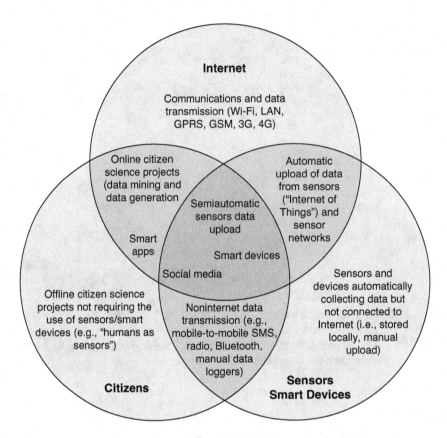

Figure 2.1 Crowdsourcing methods defined by intersections of the Internet, citizens, and sensors/smart devices (Muller et al. 2015). Reprinted with permission from John Wiley & Sons.

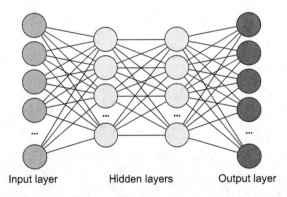

Figure 3.1 Neural network structure with input, hidden, and output layers (Marcus 2018). Reprinted with permission of author.

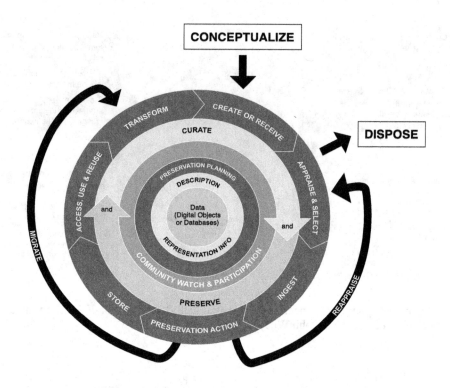

Figure 3.2 Digital Curation Centre Curation Lifecycle Model (Higgins 2008). Reprinted with permission of the Digital Curation Centre (www.dcc.ac.uk); CC-BY 4.0 (creativecommons.org/licenses/by/4.0/). Modified to render in grayscale.

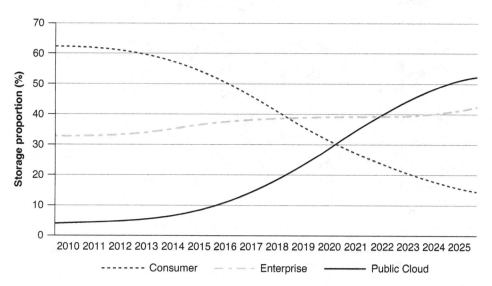

Figure 3.3 Where are big data stored? (Reinsel et al. 2018). Reprinted with permission from the International Data Corporation.

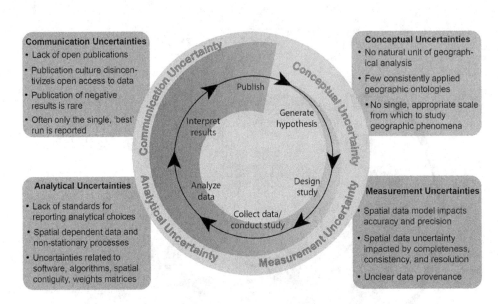

Figure 3.4 Sources of uncertainty in spatial data (Kedron et al. 2021). Reprinted with permission from John Wiley & Sons.

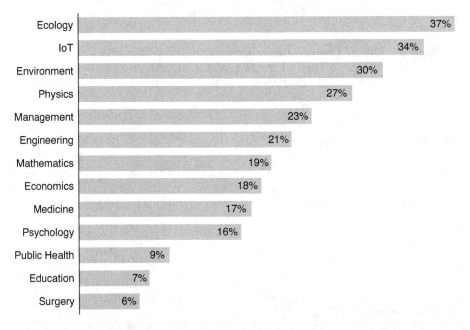

Figure 3.5 Percentage of journals with availability requirements (Klemens 2021). Reprinted with permission of author.

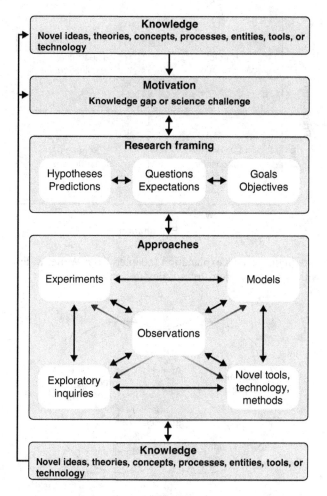

Figure 4.1 Forms of scientific inquiry (adapted from Elliott et al. 2016). Reprinted with permission from Oxford University Press.

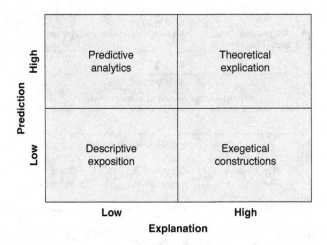

Figure 4.2 Prediction versus explanation (adapted from Waller & Fawcett 2013).
Reprinted with permission from John Wiley & Sons.

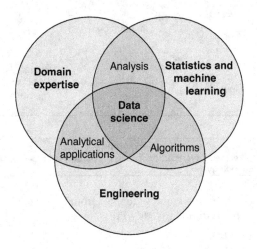

Figure 4.3 Skills needed in data science (Chang & Grady 2015).

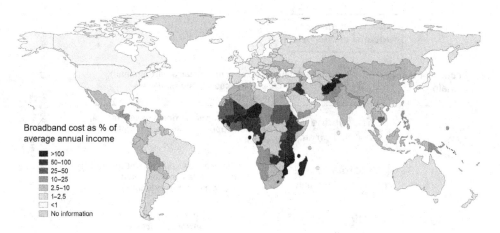

Figure 5.1 Cost of broadband subscription as a percentage of average yearly income (Graham et al. 2015). Reprinted with permission of John Wiley & Sons; CC-BY 4.0 (creativecommons.org/licenses/by/4.0/). Modified to render half of map in grayscale.

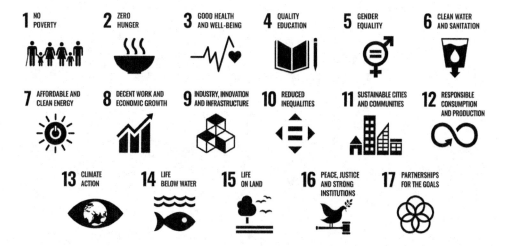

Figure 5.2 Sustainable Development Goals (United Nations 2020). Used with permission. The content of this publication has not been approved by the United Nations and does not reflect the views of the United Nations or its officials or member states.

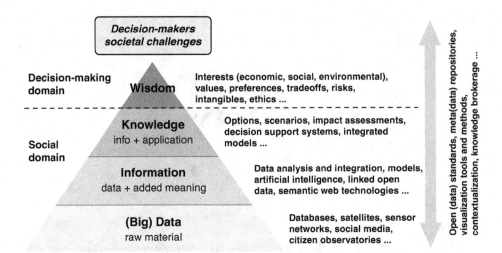

Figure 6.1 Data–information–knowledge–wisdom pyramid for human-environment challenges (adapted from Lokers et al. 2016). Reprinted with permission from Elsevier.

phenomena, such as precipitation and temperature. This autocorrelation plays havoc with the assumption that observations are independent, which underlies many standard approaches and can introduce incorrect estimates and or measures of significance (Faghmous & Kumar 2014a). Autocorrelation can also lead to incorrect data quality assessments generated by machine learning in global climate change mapping (Meyer & Pebesma 2022). Conversely, while spatial and temporal autocorrelation are typically seen as problems, they can also help with analysis. Methods that can accommodate autocorrelation can create better analyses, as when modeling temporally autoregressive terms in machine learning (Loglisci & Malerba 2017).

Human-environment data are often multiscalar. Lining up two or more data sets with varying spatiotemporal extents requires finesse. Many data science analyses draw on a single frame of analysis, such as the pixel or the individual, and rely on being able to attach data to this frame. This relationality or ability to match data sets via shared identifiers or objects (Section 1.2) is a defining characteristic of big data (Marton et al. 2013). Many questions in coupled human-environment systems necessitate accounting for decision-making at one scale (e.g., a farm field, a migrating family, animal hunting behaviors) as well as local to regional scale characteristics (e.g., cities, biomes), and even larger-scale social and ecological processes (e.g., economic shifts or climate change) (Fox et al. 2003). Scholars have to consider and capture complex relationships between structure and process while wrestling with how these interactions are *scale dependent*, changing as the scale of observation and analysis shifts (Manson 2008). One of the biggest challenges in using deep learning and other machine learning approaches is that they work well on flat, nonhierarchical data but may do less well with hierarchical or nested data sets and, by extension, real-world hierarchies.

Spatiotemporal data exhibit zonation and aggregation that can pose analysis problems. An enormous amount of data is collected at the fine scales of the household via survey, pixel-level by satellite, or individual ground-based sensors (Section 2.2). These observations are usually gathered, interpolated, or reported at a more coarse resolution for reasons related to convenience, confidentially, or analytical needs. These data are aggregated (grouped) and zoned (placed into often-arbitrary regions) in ways that pose problems to subsequent analysis. The *modifiable areal unit problem* is the general term for issues that stem from arbitrarily zoned and aggregated data (Openshaw 1983). Aggregation makes the data subject to the *ecological fallacy*, which occurs when a measure calculated for a group in aggregate can be incorrectly applied to an individual member of that group, a problem especially vexing for spatial data (Openshaw 1984). Zoning or aggregating the same underlying observations leads to different data sets that can support divergent analytical conclusions (Buzzelli 2020). For example, scholars studying human-environment interactions in health deal with this situation because they use aggregation to protect patient confidentiality and zonation to map social and environmental factors (Parenteau & Sawada 2011). Dealing with the modifiable areal unit problem in machine learning applications is gaining attention but is still a significant challenge (Zeng et al. 2020).

Multiscalar data and the systems that they represent have spatiotemporal lags because phenomena are linked and nested over varying spatial scales and time horizons. For example, climate modeling relies on grids for data storage and computing. However, analysis and modeling may require more processual data or analytical models, such as graph-based representations of physical interactions among locations or agent-based approaches to capture human decision-making. These data and models capture spatiotemporal processes, such as heat transfer among locations, that are highly dependent on local, regional, and global spatial relationships. Representing these processes traditionally requires explicit process modeling of these spatial and temporal linkages, which is antithetical to many data science approaches. Climate scientists expend tremendous amounts of time and effort teasing apart the myriad complex relationships that define a range of Earth systems. These interactions have an extraordinary range of temporal and spatial scales that span from hours to centuries and mere centimeters to the entire globe.

Data scientists and domain scholars are working to address the many challenges of model form and assumptions. For example, climate researchers use graph-based data models to link observations via a weighted network to capture relationships among locations. In the extreme, one can construct a fully connected network from a gridded database by the expedient of creating a complete matrix of weights that link every single grid cell to every other grid cell. The challenge lies in seeing whether there appear to be meaningful links among specific cells or clusters of cells by looking for correlations and anomalies, such as temperature swings in one place linking to precipitation changes in another (Tsonis et al. 2006). Such analysis can demonstrate how some regions are deeply connected, such as the tropics, where there is much apparent synchronization among cells.

Graph-based models are making their way into data-intensive modeling for human-environment research (Graafland et al. 2020). Such models typically do not capture processes but rather correlations in simple data sets presumably tied to underlying processes. For example, Kawale and others (2012) offer one of the first uses of graphs to link *dipoles*, connections between the pressure in different and distant regions to understand climate variability. They find significant relationships among the many possible candidate dipoles.

Stationarity is also an issue in the data science of human-environment systems. Focusing on the vast data volumes makes it easy to lose sight of how these data may only apply to a particular time and place. These data are often not representative of the phenomenon of interest, or they capture a dynamic process that is itself *nonstationary*, which means that the underlying process will vary over space and time. The much-vaunted success of Google Flu Trends in linking web searches to accurate predictions of seasonal flu incidence was difficult to replicate and suffered from overprediction for reasons centered on nonstationarity (Lazer et al. 2014). The underlying algorithm was overtrained and overfitted on a relatively small data set. The number of flu-related search results varied over time because of media mentions. The algorithm itself was changed repeatedly, which meant it was not stationary relative to data. More recent data science work uses location data aggregated across hundreds of millions of smartphones to forecast influenza epidemics and shows advantages over models that rely on commuter surveys to track mobility (Venkatramanan et al. 2021).

In general, many data science approaches, especially those based on artificial intelligence, are very good at interpolating within samples but less adept at extrapolating beyond data sets. Many of the early methodological advances succeeded because they worked on discrete data sets, such as sets of images linked to limited sets of labels. Coming up with out-of-sample outputs or developing more abstract models of a human-environment system is much more difficult (Lake et al. 2017). As explored in Section 2.3, there are many gaps in data regardless of source. Scholars often point to the vast data sets available in prominent research areas like global environmental change coming from satellites or long-term surface observation. However, there are many spatial, temporal, and attribute gaps in many parts of the world (Pahl-Wostl et al. 2013). A continuing focus for the data science of human-environment systems is using approaches like machine learning to help remedy these gaps and deal with them during analysis.

4.2 Getting It Wrong, Getting It Right

Data scientists and domain experts need to work together to get human-environment systems right. Computer scientists, statisticians, physicists, and other data specialists do a good deal of data science on various human and environmental systems. Much work is done by researchers, "most of whom have no firsthand experience of producing the experimental data they are analyzing. This has contributed to an exaggerated trust in the quality and comparability of the data and to many irreproducible results" (Strasser 2012, p. 86). At the same time, domain scientists have lagged behind data scientists in leveraging promising advances in collecting, handling, and analyzing large data sets. They also run the risk of wanting to dismiss data science approaches outright because they are being carried out by people perceived as scholarly outsiders.

Consider work in applying big data to health to advance human well-being. This research has led to significant increases in false reports of correlations between potential causes and disease outcomes. The primary strategy for dealing with false correlations is using large-scale replication studies, but these tend to be difficult or impossible in many contexts. The way forward is to invest in data science with a stronger epidemiological foundation.

Big Data analysis is currently largely based on convenient samples of people or information available on the Internet. When associations are probed between perfectly measured data (e.g., a genome sequence) and poorly measured data (e.g., administrative claims health data), research accuracy is dictated by the weakest link. Big data are observational and are fraught with many biases such as selection, confounding variables, and lack of generalizability. Big Data analysis may be embedded in epidemiologically well-characterized and representative populations.   *(Khoury & Ioannidis 2014, p. 1054)*

The primary challenge for data scientists is their often limited understanding of substantive domains, while experts in these domains are not always comfortable with data science approaches. Data scientists can be excellent researchers yet have little training or expertise in the subject matter and do not often attempt to engage with domain experts in the underlying substantive research areas. In other cases, data scientists will explicitly reject engaging with the underlying domain to minimize bias in model construction and subsequent analysis. This approach has the benefit of bringing fresh eyes to problems, but there is likely a limit to its effectiveness over the long run. One result of this narrowness is what some researchers call "data alchemy" and insufficient scientific exploration of data science fundamentals, including a focus on replicability and theory development. These shortcomings result in a mix of outstanding scholarship and research that is often interesting and statistically sophisticated but fundamentally wrong or trivial in the subject matter (Henke et al. 2016). The way forward for many people is the potential for better collaborations between data science and domain fields and a greater focus on the science of data science.

4.2.1 Data Alchemy

A primary critique of data science methods is that they can often predict but cannot explain. Data science is likened to a digital microscope that allows us to find new patterns and create predictions. However, consideration of processual explanations is absent in much of this work. There is genuine worry that the scientific metaphor for big data is not the microscope or telescope but alchemy marked by a near mysticism. Researchers across many fields see how deep learning and cognate methods are tremendously effective. They also acknowledge that many features of data science methods, like artificial intelligence and deep learning, are poorly understood (Sejnowski 2020).

Some data scientists argue that the field relies too much on a mix of ritual and blind trial and error in performing the most fundamental aspects of research. In the words of artificial intelligence expert François Chollet, "'People gravitate around cargo-cult practices,' relying on 'folklore and magic spells'" (Hutson 2018, p. 861). Thousands of papers and presentations are based on informal and often untested sets of tools and approaches for handling the most fundamental facets of artificial intelligence, such as training neural networks.

Is artificial intelligence the new alchemy? That is, are the powerful algorithms that control so much of our lives – from internet searches to social media feeds – the modern equivalent of turning lead into gold? Moreover: Would that be such a bad thing? According to the prominent AI researcher Ali Rahimi and others, today's fashionable neural networks and deep learning techniques are based on a collection of tricks, topped with a good dash of optimism, rather than systematic analysis. Modern engineers, the thinking goes, assemble their codes with the same wishful thinking and misunderstanding that the ancient alchemists had when mixing their magic potions. *(Dijkgraaf 2021, p. 1)*

Data science may spend too much time on a small set of established problems. Much of the appeal of data science and a source of evident success lies in its use of *common task frameworks* where different researchers compete to solve a well-defined problem against a single shared data set (Donoho 2017). This common task framework approach has its origin in *machine translation*, the use of computers to translate among languages. Language translation is a good research problem for this approach because it offers some ambiguity since translation is both an art and science but is relatively narrow in scope. Data science struggles with empirical rigor and replication issues because scholars focus on squeezing out incremental increases on well-understood problems instead of better understanding underlying computational processes (Sculley et al. 2018). Watts (2017) calls for extensions of the common task framework into the social sciences, which would apply to human-environment scholarship. However, this approach may work best for mesolevel problems that are neither so complex as to defy analysis nor those that are trivial.

Data science could use more science to correct missteps in machine learning, deep learning, and other core approaches. Growing numbers of data scientists and others call for more scientific inquiry, including the hallmarks of experimentation, hypothesis testing, and replicability (Mannarswamy & Roy 2018). For example, while deep learning approaches have some ability to deal with new data or novel situations, in general, they are contingent on the data against which they are trained. There is a need for purposeful exploration of how brittle these methods can be when confronting new data or a change in circumstances and finding ways to make machine learning more human in its ability to continue to learn after initial training

(Chen & Liu 2016). Under this vision, data science could be more scientific in seeking to develop theory, with a process of small steps as part of playful but rigorous experimentation (Dijkgraaf 2021).

There is also growing demand for *explainable artificial intelligence* (XAI). This approach couples methods like machine learning with greater interpretability and explanation, especially for opaque deep learning results (Samek et al. 2021). Many of these approaches to interpretation are based on established mathematical and statistical methods for decomposing and visualizing results. Several fields are leading the way, including military research and medicine, because the stakes are high when people's lives are involved (Knight 2017). The XAI approach also builds on the work examined in Section 4.2.2, where researchers engage in boundary-crossing work, build hybrid models, and build processual elements into models while using established forms of sensitivity analysis to see how changing inputs can lead to different outputs (Voosen 2017).

Another way to make data science more scientific is to address issues around reproducibility. Reproducibility is a critical challenge in many scientific fields (Section 3.4), and artificial intelligence research has its own problems. A survey of 400 artificial intelligence papers presented at major computer science conferences found that just 6 percent of these papers included code for the algorithms being written about, and only a fifth to a third of the papers offered basic features of reproducibility such as test data or code of the variables (Gundersen & Kjensmo 2018). In response to these issues, several institutions and research organizations are launching platforms for reproducible experiments (Isdahl & Gundersen 2019).

However, such efforts will likely share in many of the problems that reproducibility efforts face. At the heart of the problem are many disincentives and few incentives for reproducibility.

AI researchers say the incentives are still not aligned with reproducibility. They don't have time to test algorithms under every condition, or the space in articles to document every hyperparameter they tried. They feel pressure to publish quickly, given that many papers are posted online to arXiv every day without peer review. And many are reluctant to report failed replications.   *(Matthew 2018, p. 726)*

Reproducibility is essential for robust data science, but the field is like many others in needing improvement.

4.2.2 *Getting It Right by Crossing Boundaries*

One way forward is investing in boundary-crossing work. These efforts take many forms. One is theory-guided data science, where data science models include representations of theoretical concepts or practical rules to guide model development. This form of data science uses physical principles or other theoretical concepts to guide model development (Karpatne et al. 2017). This approach is gaining currency primarily in the environment and physical sciences, including climate science, hydrology, fluid dynamics, and quantum chemistry. These fields have large data sets and can express core physical principles mathematically in ways that are well-suited to modeling. These mathematical functions and thresholds can be used to constrain the model forms, variables, or parameters used by artificial intelligence and machine learning methods to make them more realistic. There is also a focus on using domain knowledge to help interpret or refine model outcomes or develop hybrid models that combine data-led and theory-led conceptualizations of modeled systems. Examples include using machine learning to impute missing information about how plant traits vary with environmental conditions (Schrodt et al. 2015) or using topography to classify land cover and water body boundaries in remotely sensed imagery (Karpatne et al. 2017).

An example of theory-guided data science is data mining to sift through vast amounts of climate data. Even when using these methods fit observations well, their predictions may be physically inconsistent or even implausible because they ignore real-world constraints. Encoding physical principles and rules into machine learning approaches can bound the search space or predictions that these methods provide (Klocke et al. 2011). This approach involves building physical principles into models to limit the relationships and extremes one could find in these data, such as physics-guided data mining (Ganguly et al. 2014).

The focus of this work is to characterize better extreme climate events, assess uncertainties, and enhance climate projections. There is also a need to understand better the underlying dynamics, such as carbon dioxide production or the effects of sudden shifts in ocean circulation. Much of this work centers on applying fairly robust methods, such as dimensionality reduction and clustering, to model effects on carbon dioxide emissions from energy consumption and economic activity (Mardani et al. 2020). Related work centers on using dimensionality reduction and machine learning to downscale and model climate elements such as precipitation (Sarhadi et al. 2017).

Conversely, although data science allows scholars to ask new questions and develop novel insights, too many domain-centered researchers are dismissive of – or actively hostile to – perceived interlopers armed with new approaches and ideas. These protectionist attitudes limit scholarly conversation and engagement around when big data works and when it does not. It is incumbent on domain scientists to step up and engage with data scientists. For example, much of big data

is social data – that is, data about the interactions of people: how they communicate, how they form relationships, how they come into conflict, and how they shape their future interactions through political and economic institutions. It is the responsibility of social scientists to assume their central place in the world of big data, to shape the questions we ask of big data, and to characterize what does and does not make for a convincing answer.

(Monroe et al. 2015, p. 71)

The increased sophistication of pairing data science and domain-specific approaches is vital to human-environment research. The primary value of big data approaches may lie in their use as an inductive, exploratory tool to highlight potential correlations during the preliminary phases of research. Data science offers the potential to design new and better experiments, especially on subpopulations that are ignored by much of the current work in quantitative scholarship or to leverage new data sources that can answer new questions. Returning to the climate change example, deep learning methods may resolve computational challenges in global climate modeling, such as helping capture fundamental moisture–cloud dynamics that are not modeled well (Gentine et al. 2018). Machine learning models can also be used as proxies by approximating full-fledged climate models (Reichstein et al. 2019). This approach saves the processing effort and the time needed to do an entire run of a climate model at the cost of some accuracy. Similarly, recent work in mathematics points to how deep learning methods can aid human intuition, pointing out how these methods can provide new insights into existing problems or help bridge gaps in knowledge (Stump 2021).

Bridging scholarly domains also implies mixing methods. At some points in the data science workflow, researchers find it helpful or necessary to employ other cutting-edge or domain-appropriate approaches such as network analysis or multi-level regression to see which initial correlations are accurate or useful (Salmond et al. 2017). Computer and statistical modeling, in particular, are increasingly used to bridge the inductive insights of big data and the deeper analysis afforded by domain-specific methods. For example, geomorphology looks to machine learning and points to how

the development of models that are scalable and can be translated between sites is dependent on experience in the field. Although [machine learning] models are shown to be effective as a surrogate to process-based numerical models, they are only as good as our conceptual understanding of landform and landscape form and evolution. This means that [machine learning] is simply a new and powerful tool in the proverbial belt of the geomorphologist and should not come at the expense of the field tradition that informs us of whether [machine learning] results are accurate, transferable, and scalable.   *(Houser et al. 2022, p. 1260)*

Data science initiatives on coupled human-environment systems, when done well, are essentially *eScience projects*, interdisciplinary science projects that involve multiple researchers from across disciplines who collaborate with information technology experts. Pennington (2011) notes that the breadth and depth of disciplinary differences among these researchers can be so great as to complicate the development of an integrated conceptual framework. Nevertheless, this framework is necessary for joining data and methods to drive science. Data or methods absent a conceptual grounding are necessarily stunted, but establishing a shared conceptual framework is usually a roadblock to collaborative eScience efforts. A cohesive data science (in the sense of science about how data is used) should include an explicit focus on how big data projects can productively develop shared data, methods, and theories. This sharing involves developing concepts in common, including their own terminology. When dealing with data and modeling, there is also a call for integrated data, tools, and methods at the heart of data science.

Integrated projects can help bridge divergence in approach or interests between computational researchers and domain scholars. Data science sometimes is a solution in search of a problem. A case in point is the EarthCube, which aims to advance geoscientific research by integrating data sets, models, and attendant software to support the discovery and reuse of data. This integration is essential because it can help scientists tackle complicated human-environment issues, including global climate change, severe weather impacts, and natural resource management. EarthCube has been very successful.

However, integration is a process and not a single outcome. Surveys of actual and potential users highlighted potential barriers to integration. They raise questions about the actual need for this kind of infrastructure, the lack of support for domain scientists who want to engage with it, and the need for better balance among the research communities, particularly between domain scholars and computer scientists (Cutcher-Gershenfeld et al. 2016). Carefully trained scientists with deep knowledge in domain areas must work with experts in causal methods, especially when dealing with coupled human-environment systems.

4.3 Data Big and Small

Linking small and big data is an essential boundary-crossing activity for successful data science. Small data have great value in and of themselves and can bring greater value to big data, particularly advancing theory. In using the term small data, scholars are not simply looking for methods required to make big data tractable to humans by subsetting data or developing practical analytical methods that abstract meaning or models from sprawling data sets (Sieber & Tenney 2018). Definitions of small data suffer when these data are treated as big data subsidiaries or simply as a way to provide value to data science. While data science is sometimes seen as diminishing the role of theory development driven by qualitative data and other kinds of more traditional, smaller data sets, it is unlikely to have that effect in the long run. There are many successful instances in which data scientists worked with domain scientists to understand better the nature of fieldwork and the potential to combine it with big data. At its simplest, domain experts and small data serve to ground big data, and small data are valuable on their own (Hampton et al. 2013).

Integrated human-environmental modeling is one way to link small data with big data. A wide array of model development, calibration, and parameterization is driven by stakeholder participation via qualitative interviews and facilitated work-shops (Voinov & Bousquet 2010). Of course, while these workshops can be extraordinarily helpful in model specification, they are resource-intensive, given the need for trained facilitators and the time and effort of participants. They require a good deal of time, training, and sharing of experience to develop the necessary trust for fruitful collaboration (Adams 2014). Also, turning qualitative insights into quantitative data to drive model development is a difficult task that requires a good deal of expertise in qualitative methods generally and, by definition, experience with specific substantive domains.

It is also challenging to balance depth of insight for a specific time and place against the need to generalize as part of the larger scientific enterprise. Experts can also use statistical methods to link their knowledge and experience to data science analyses to provide demonstrable improvements in prediction (Kreinovich & Ouncharoen 2015). There is also the promise of using big data tools, such as shared data sets and cyberinfrastructure, to help combine small data sets into larger ones more amenable to data science approaches while still preserving the benefits of small data sets (Kitchin & Lauriault 2015).

Small data offer nuance and can elucidate details that are glossed over in big data. As discussed elsewhere in this volume, there is much evidence that big data can encode bias in ways that matter to, or disproportionately affect, smaller communi-ties, be they rare flora or a population minority. Kientz (2019) terms these N-of-1 studies and demonstrates how they can yield profound insights into particular people, places, or problems. These studies can prove helpful when encountering new phenomena, like a member of a new species, helping frame a case study that invites deep investigation, or addressing the lessons from rare events such as accidents or disasters (Prior 2016). Small can also mean geographically bounded, as when using detailed census data tied to spatial boundaries and collected and disseminated with processes that build quality control and extensive documentation (Goodchild 2013). In working with small data, there is an emphasis on the researcher practicing reflexivity and being attuned to bias in data interpretation. For many fields, advanced knowledge is born of intensive study of small areas: a single neighborhood, a stand of trees, or a meters-long transect of the seafloor.

Manovich (2011) distinguishes between surface and deep data in many discip-lines. *Surface data* deal with large extents and many potential variables, such as a social scientist using census data or a climatologist using global data with one-degree pixels with dozens of variables. *Deep data* drill down to specific instances, such as a social scientist engaging with a single survey respondent over the years or an ecologist examining microclimates over meters to understand plant growth. One way forward with the data science of human-environment systems is to bridge the surface and the deep. Data science can help the social scientist follow many individuals over the years, while climatologists can use the growing array of measures collected by millions of mobile devices.

There are complications with this vision to integrate data small and large. As seen in any number of cases, data are censored, biased, and incomplete in ways that affect their interoperability at the best of times. Insights from deep work, like ethnography or field biology, differ from those granted by the automatic processing of massive data sets. Both approaches may convey nuance, but it is very likely to be different kinds of nuance, especially given how easily flummoxed are even the most advanced forms of machine learning or natural language processing (Manovich 2011). Finally, big data is still a domain made up of specialized fields, requiring an exhaustive list of skills and aptitudes that can usually only be found in interdisciplinary teams. Small data are exemplified by long-term field observations, case studies, or participant observation. Ethnographers, biologists, ecologists, and geologists are among the many researchers who draw on long traditions of careful scholarship that entail close study of their object of study. These scholars may face pressure when bigness is desired by funding agencies or become more central in scientific practice and norms. Combatting these potential inequalities can take several paths, including better linking small- and large-scale studies alongside more mixed-methods research, refining theory to winnow down uncertainties, and pooling data to create larger data sets. Many of these steps are helped by institutional support and infrastructure (Section 6.3).

4.3.1 Environmental Sciences

Environmental science exemplifies the tensions and possibilities of big data versus small data. Extraordinary work in ecology often depends on small-scale field studies. While there is a great deal of interest in, and potential for, networks of embedded ecological sensors and remote imaging, the vast majority of species occurrence data is still field-based. Science is often undertaken by individual researchers or small teams over small spatial and temporal scales, and moving up scales can meet with resistance. For example, the International Biological Program (IBP) was an early attempt at large-scale international research. It had some success but also met pushback from many scholars who saw it as unnecessarily intrusive and even antithetical to the nature of fieldwork (Aronova et al. 2010). For many ecologists and other scientists working in the environment, the most common mode of inquiry is to work in a single area, often on the order of tens to hundreds of meters, to sample plants, animals, and soil and then publish on this area while linking to a broader scholarly conversation.

Bridging small data to big data is helped by data infrastructure. The US National Center for Ecological Analysis and Synthesis (NCEAS), for example, encourages and facilitates new research with existing ecological data, especially small-site data. This work necessitates a system where individual field sites and their data could be shared and aggregated. Like other science infrastructures, the NCEAS owes its existence to a blend of bottom-up organization and large infusions of funding from the US National Science Foundation (NSF). It also requires dedicated funding for staffing and visiting fellows, postdoctoral fellows to help build the next generation of scholars, and working groups to tackle issues and substantive research questions as they arise. Workgroups can tackle seemingly small yet ultimately important issues, such as handling and storing data on biomass for species with complicated seasonal variation, requiring discussion, argument, and communication in real-time with field researchers (Hackett et al. 2008).

Some infrastructures support deep-but-narrow data-intensive enterprises. These are projects like the Avian Knowledge Network (AKN), which contains tens of millions of bird observations across hundreds of variables for more than 400,000 locations in the United States (Kelling et al. 2009). Another example is the Flora of North America Project (FNAP). This effort sought to identify and catalog all of the native plant species in North America, and was among the first of such systems to use the Internet to support a community of scholars who would develop and share information for research and teaching (Spasser 2000). The Organization of Biological Field Stations (OBFS) is a system of field stations, some existing for over a century, that engages in primary research alongside education and outreach. There has been an increased focus on developing essential cyberinfrastructure to support work within and among these stations (McNulty et al. 2017).

The advent of the Long Term Ecological Research Network (LTER), funded by the NSF, opened new vistas by developing six (and eventually over two dozen) sites for intensive and long-term ecological research. It was designed to span a range of biomes and climates and facilitate the time scales necessary to answer ecological questions (Magnuson & Bowser 1990; Hobbie et al. 2003). From its earliest days, the LTER focused on data sharing in the form of the LTER Network Information System (NIS), a cooperative, federated database system. The LTER ecological work embraced the study of evolution and long-term change. This project also prioritized data collection when it was often considered an afterthought, scaling up site-specific data collection to multisite information systems and practices shepherded by statisticians and others trained in data instead of ecologists or modelers (Baker et al. 2000).

Building a culture of sharing would help bridge the worlds of small data and big data. Ecology is likely akin to other fields in having a range of data-sharing practices (Section 3.4). Hampton and others (2013) examined a broad sample of environmental biology projects funded by the NSF in the 2000s. They found that 57 percent shared no information, 15 percent shared some, and only 28 percent shared all of their data. Of those that shared their data, the vast majority (81 percent) of projects were of one specific kind relating to genetic data shared through two specific databases. This rate of data deposition was driven in part by expectations in the field that these data should be shared as part of the publishing process early on, creating a culture of sharing and reuse. There is even more room to develop expectations around regularly and automatically contributing to community databases to realize the potential of big data.

Small data need not become big data to help data science. Fieldwork is complicated because the world is messy, and from this complication comes the N-of-1 studies and their insights. Even scholars in fields with agreed-upon standards routinely confront local variations and personal contexts that encourage new kinds of study and analysis. These issues make work creative and challenging as researchers adapt to changing circumstances driven by shifts in weather, socioeconomics, and politics (Kohler 2002). Reconciling the challenging nature of local, grounded research with mesolevel instrumentation and field-site specific designs and macrolevel intersite networking and data sharing requires iterative improvement, goodwill, and communication. Successful efforts may require making sensors more mobile or deployable for months versus years in fixed emplacements, allowing researchers to change locations in response to on-ground research needs. It should also leave room for repairs, changes, and improvements in the field. Much of this boils down to the idea that doing measurements well requires human oversight and often active intervention to make sense of the world (Mayernik et al. 2013).

Embracing a mix of small and big data in environmental work is vital because both kinds of data act as shadow selves of people or the environment and are acted on in ways that manifest in reality. Scholars have long surveyed people and the environment in ways that produce feedback for individuals, society, and the environment. Fisheries management, for example, involves talking to a range of humans and assessing the environment in many ways and translating this information into action. The challenge is that this work is usually hard to do well, and it is not clear how big data will help when many of the challenges are integrating data into management systems. However, there is promise in using genomics and other data science approaches to make these resources more sustainable (Lee & O'Malley 2020). There are related efforts in agriculture, where farmers and others on the ground want to access information that is tailored to their biophysical and socioeconomic context while also having some insight into how valid the models and underlying theories are to their circumstances (Antle et al. 2017). Efforts such as the Agricultural Model Intercomparison and Improvement Project (AgMIP) are oriented toward reconciling differences among models. Such projects increasingly look to data science to integrate and extend modeling efforts and aid theory development (Lokers et al. 2016).

4.3.2 Environmental Humanities

The environmental humanities and the digital humanities offer a critical take on the tension around data science, bridging big and small data, and what data mean for theory. The digital humanities blend data science with the arts and humanities, including art, music, languages, history, and philosophy. Digital environmental humanities examine human-environment interactions, especially the digital revolution, global climate change, and attendant social, economic, and political conflict (Travis & Holm 2016).

Scholars range from gleeful advocacy (or outright zealotry) of data science to a more measured wait-and-see attitude to outright skepticism, especially regarding social aspects of human-environment systems. In the words of one scholar,

do Big Data and neural computers tell us much about society? To what extent can the most powerful neural network imaginable, which possesses all the Big Data in the world, really grapple with human desires, with issues of justice, and with deliberative and conflictual processes where there is no correct answer, no rabbit to pull out of a hat, but clashes of will, persuasion, emotions, life and death? Furthermore, apart from learning about these things second-hand, what can a computer – even one that has taught itself to learn from Big Data – really know about fear, joy, hate or even, more prosaically, about lying in the grass with the sun shining on its face. *(Shearmur 2015, p. 966)*

Data science holds appeal for humanistic environmental inquiry. A significant focus in the field is the movement to digitize vast paper holdings of books, images, and other records. This work is an extraordinary achievement and a great opportunity in the eyes of some scholars. To go from being able to read (and think about) dozens of volumes a year to having access to measures derived from literally millions of books is a boon for some kinds of research. Concerning human-environment systems, in particular, there is scholarship on how the digital humanities interfaces with, and advances, work on the environment. This humanistic inquiry meets the ongoing need to make sense of environmental change and how humans can preserve and memorialize what is being lost to this change. Humanistic thinking is also central to understanding the ethical and moral dimensions for action and struggle or coming to accept some aspects of a world beset by myriad human-environment calamities (Nowviskie 2015; Ladino 2018).

While early digital humanities work focuses on simple descriptive statistics, such as tracking the rise and fall of specific terms in novels, data science tools enable more sophisticated analysis. Boosters see this digital turn as bringing more rigor to a field that uses approaches such as *close reading*, the deep interpretation of texts leading to analysis and synthesis undergirded by literary and critical theory. Some of the earliest work in this realm argues that scholars in the literary or critical vein would be well served by using data science tools, including counting, graphing, and mapping, to get beyond the narrow confines of limited reading. Moretti (2005) argues that literary studies have been random, unsystematic, and overly driven by idiosyncratic personal views or being beholden to a small subset of works at any one time out of hundreds of thousands of available works. In contrast, *distant reading* occurs when the entirety of world literatures can be encompassed as a way to move beyond the focus on Western European novels that characterizes much of humanities scholarship and embrace a much broader array of genres and people. Distant reading furthers the idea that literary history and concepts such as aesthetics are much more complex and different than commonly understood by many humanities disciplines that spring from narrow and Eurocentric bodies of source material (Moretti 2000).

Critics of the digital humanities evoke many of the issues that bedevil data science more generally. There is the worry that this work is superficial and misses the point of specific texts. Scholars point to the power of spending an entire career focusing on one or two works, reading, rereading, and interpreting them with new lenses. This work requires deep knowledge of the author's life, times, and milieu. In this view, replacing deep and engaged analysis with a thin examination of thousands of texts misses the point of literature and other human objects, given their depth and complexity, and needlessly simplifies and caricatures what close reading involves. This approach is also blind to extensive domain knowledge – understanding a single work of fiction can require intimate knowledge of the writer and the broader social, cultural, and environmental context (Smith 2016).

Is there a middle ground in using data science in environmental humanities research? A potentially helpful perspective on digital humanities is that it can play a significant role in analyzing data science. This view supports the idea that the field of data science should devote effort to science about data (Section 1.3). In this case, digital humanities have a lot to say about the trappings of information writ large, including software, hardware, and their social contexts. The humanities have led the way in understanding how social, political, and economic power guides the nature of scholarship. In addition, the culture of the digital environmental human-ities very much promotes critical thinking and public engagement around the social dimensions of nature. These qualities can only aid data science for human-environment work (Posthumus et al. 2018).

The humanities also have much to say about how most scholarship today, even the most avowed humanities-oriented research, takes place in a rich soup of digital media. There are tradeoffs. Gaining digital access to vast numbers of journals and articles has changed the nature of discovery. Most digital portals now have a recommendation system that suggests other articles based on the one being read. However, it is hard to replicate the experience of being in the library stacks and the thrill of discovery a person encounters when searching one book on shelves and finding by happenstance another that is even better or opens up new lines of inquiry.

Digital humanities demonstrate that reading and writing are forms of embodied cognition that combine automatic, learned, and controlled mental processes tied to the media in question (Elfenbein 2020). Reading on a screen or writing on a computer instead of reading a book or writing notes in longhand are cognitively different. So too, there must be differences in how a biologist interacts with the world when writing field notes in a paper journal or composing images with a camera versus examining reading captured by remote monitoring stations or images taken with artificially intelligent photograph traps.

Finally, data can be richer and more meaningful when researchers and the lay public dive into the data to understand their benefits and challenges. It is important to situate the data in the lives of the people collecting it (Gabrys et al. 2016). This effort can also lead these data to be better combined with other data sets. In particular, iterative engagement with data in many forms can weave together raw observations, mixes of text and imagery, and visualizations and analyses via logbooks, sensors, online data access, and mapping. Citizen scientists can move deeper into the data and explore the data, potential problems, blind spots, and cofounding variables or issues (e.g., various sources of noise or pollution). There is a helpful tension in not seeing stark differences between data and reality, or virtual and the real, because many phenomena can take on aspects of both (Boellstorff et al. 2016). The digital repre-sentation of something can be acted on in ways that affect the entity in reality, while the phenomena in reality will eventually impinge upon data.

4.3.3 Computational Social Science

Computational social science builds on decades of quantitative research and has embraced data science and other forms of data-intensive inquiry. This work uses various mathematical, statistical, and computational tools to understand human systems (Lazer et al. 2009). The social sciences are home to many debates about balancing quantitative and qualitative work. These arguments are usually proxies for deeper divisions in epistemological stance. One advantage of these conversations is that many computational social scientists are open to data science critiques while simultaneously seeking to advance these approaches. They are also leading the way in integrating qualitative and small-data work into data science and computational social science.

Computational social scientists often focus on the importance of theory in guiding data science. Theory can

help discriminate noise from signal, and provide the right context for that interpretation. This is particularly important when so many of the models that are being applied to social systems were developed by mathematicians, physicists, or computer scientists to understand the behavior of other systems that have no agency – that is, no actors capable of processing their own information about the world they inhabit and that are free to react in accordance. True, some aspects might be generalizable, but these tend not to be the interesting ones, at least for a social science audience.   *(González-Bailón 2013, p. 154)*

For example, much work in using computational approaches in human-environment scholarship centers on sentiment mapping of tweets and other social media, which may be great at establishing spatiotemporal patterning (bearing in mind the caveats of using such data; Section 2.2.6). However, it may not be especially good at capturing the processual why questions without more profound engagement with social science theory. Social scientists engaged with the shortcomings of data science from the beginning, noting the potential for traditional domain expertise being lost or ignored as data science weighed in on social phenomena (Savage & Burrows 2007).

Many social sciences have experienced long-running debates about quantitative research, and these conversations bear on data science. The most ardent critics of big data point to the various challenges of quantitative analysis in general. For all of its power, quantitative work can overly simplify the complexity of real-world processes. It can be too mechanistic and ignore vital facets of lived experience not typically amenable to quantification, such as meaning, ethics, or moral aspects. Data science may not be a great way to do some kinds of social science. For example, the Fragile Families and Child Wellbeing Study is a high-profile effort that collected data on over 4,000 families across many variables. Project investigators invited other teams of researchers to use a range of methods to make predictions about life outcomes, such as school performance, for the families and individuals in the study (Salganik et al. 2020). This comparative effort suggested that machine learning methods were ineffective in predicting many social variables. Of course, there are limits to the generalizability of these findings. The study data are not as voluminous as many big data proponents would consider big (see Section 1.2), even though they offer variety and value. This study is extensive for social science, but it may not be large enough for data science tools to demonstrate their prowess. However, the point Salganik and others make is germane to the bigger picture of data science.

Many processes are complex, and machine learning is not necessarily going to do better than simpler models because the underlying processes are complex in ways that belie apparent patterns in these data sets. Related is the idea that prediction is separate from, and often less helpful than, description or causal inference, especially in the face of uncertainty and error (Section 4.1.2). Like most good work in the social sciences, machine learning emphasizes model validation and careful assessment of model parsimony when used well. It helps to use several approaches to assess the lower limit on predictability (the point where noise takes over) and then choose a suitable method for hypothesis testing and theory development. At the same time, social scientists are arguably too wary of overfitting models, p-hacking, or data-dredging when experimenting with data science approaches (Hindman 2015). There is room to embrace some of the more inductive elements of data science without losing sight of the larger drive to theory development.

Similar dynamics are at work in the idea of data science helping generate hypotheses. Methods like machine learning may be used to sift through data to find relationships and, from these, hypotheses. An example is the application of machine learning to a large social science data set, the World Values Survey, which has about 300,000 observations for close to 1,000 variables. The investigators argue that this approach is better than many standard regression methods in determining nonlinearities and finding correlations. "This research is likely to be of interest to all researchers in the social-behavioral sciences who work on hypothesis testing because it demonstrates a general method to generate novel hypotheses using machine-learning techniques. This method can be applied in any field in which researchers have access to reasonably large datasets" (Shetal et al. 2020, p. 1223). This work points to a future where machine learning may be used to inductively trawl through large data sets to point to possible relationships and hypotheses. Of course, for this future to happen, this work will have to account for the various challenges posed by sample issues (Section 4.1.3) and model form (Section 4.1.4).

Social scientists are used to balancing various methods and kinds of data and have embraced the role of qualitative work in understanding data science lessons. Consider anthropologists who study relationships between humans and the environment. These researchers use qualitative research methods that include ethnographic fieldwork, interviews, and focus groups to understand better how humans make decisions. These scholars find these approaches offer flexibility and the ability to quickly change focus as dictated by the circumstance (Yang & Gilbert 2008). Qualitative research methods are a powerful way to collect and curate data and, more importantly, capture something that big data usually do not, namely understanding human decision-making processes and insight into underlying motivations (e.g., Spies et al. 2017).

Ethnographers argue that they are well-positioned for working with big data because they are used to large, sprawling, and indeterminate systems: "ethnographers are less daunted by the large haystacks that other researchers find so troubling. To do fieldwork is to live with sensorial excess. Our haystacks are always 'too large'" (Knox & Nafus 2018, p. 14). They argue further that the ethnographical practice of iteratively exploring data is a way to develop *plausible coherence* when a narrative is compatible with experience and theory. This view is compatible with the desire to combine multiple forms of evidence and knowledge in understanding human-environment systems. It also points to the value of qualitative interrogation of data and models to establish trustworthiness (Section 2.3.1).

Many social scientists take a wait-and-see attitude toward data science. The primary argument for a tentative embrace of data science is that it uses many existing methods but applies them to larger data sets (Clark & Golder 2015). There is much potential for keeping existing approaches and scaling them up to work with big data. There is also emerging evidence that data science can help develop better models by identifying promising correlations and helping choose among models (Athey 2019). For many social scientists, the overriding goal of scholarship remains to develop well-specified models with a strong foundation in theoretical antecedents, which allows them carefully to trace causation in terms of social structures and individual decision-making (Brady 2019). As data science matures, it will hopefully continue to broaden relationships with other fields, prove its worth in advancing theories, and deal with critiques in ways that encourage social scientists to engage with the big data of human-environment systems.

4.4 Data Science Education

Conversations around theory development in data science overlap with data science training and education. How students are taught data science reflects and will continue to mold the way people think about theory's importance (or lack thereof). Data science has roots extending decades and has evolved from data-intensive science in fields like biology or physics. However, much recent work is dominated by computer science and engineering, and statistics and mathematics to a lesser extent. There are increasing calls to broaden the remit of data science in education to embrace better the many domains that use and advance many facets of data science. Much of the work articulating the nature of data science education is being taken up by data science programs and academic institutions, alongside efforts in the private sector and professional organizations.

The US National Research Council hosted a workshop in 2014 on training in big data. Participants agreed that training in big data, data science, and data analytics could span multiple disciplines but should have at their core computer science, machine learning, statistics, and mathematics (National Research Council [NRC] 2015). There is a nod to domain science, but it is secondary to these core fields. Given the looming importance of big data and data science, the British Academy, a society focused on the social sciences and humanities, bemoaned the quantitative skills deficit in these fields. It advocated for a more expansive vision of data science.

We have an urgent need for methodologists able to break new ground in statistical techniques, with a clear mission to develop methods for use with social science data; informaticians, able to manipulate and manage data from a single large source or multiple sources; social scientists grounded in a particular discipline, able to undertake detailed analyses of large and complex data sets; and social scientists able to interpret and understand statistical results and statistically based argument.   *(British Academy 2012, p. 5)*

Another early effort, the US National Institute of Standards and Technology Big Data Interoperability Framework (Chang & Grady 2015), pointed out how data science lies at the intersection of statistics and machine learning, engineering, and domain expertise and therefore draws on a variety of skills and aptitudes (Figure 4.3).

Recent articulations of data science education argue for moving beyond a focus on computer science or engineering and embracing statistics and domain knowledge, although many programs still fall short. The report *Data Science for Undergraduates: Opportunities and Options* (NASEM 2018) extended the idea

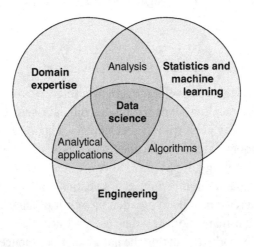

Figure 4.3 Skills needed in data science (Chang & Grady 2015).

of core academic areas to the primary goal of helping students gain data acumen. This primary goal is tied to a series of secondary goals, including incorporating real-world applications; enhancing communication and teamwork skills; considering ethics; and emphasizing diversity, inclusion, and increased participation. The primary goal of acumen is relatively narrowly defined. It focuses on combining existing programs into a single workflow; ingesting, cleaning, and wrangling data; considering data issues; assessing sound analytical methods; and communicating effectively. It remains to be seen how well the average data science program can help students achieve data acumen while also equipping them to handle more extensive goals.

There is room for more balance among data science fields in how scholars talk about what is taught and why. Some statisticians express frustration with how data science is taught in many institutions. They argue that statistics covers much of what data science entails, but data science programs per se may not teach these topics well or in sufficient depth (Peng 2017). Modern statistics concerns itself with vital aspects of data science: data exploration and transformation; the mix of computing, modeling, and visualization; and the science of data science. Some of these topics are difficult to teach systematically, especially because learning the basics of statistics is already time-consuming and challenging. There is a good deal of data plumbing and scut work involved in working with big data. In response, people are writing statistically minded texts addressing the job of working with data in packages like R or python (e.g., Wickham & Grolemund 2016). Another critical challenge is that many statistics curricula assume profound numeracy on the part of students, to the point where the field " has been blinded to recent developments and has missed an important opportunity to cultivate more academic leaders in this area" (Peng 2017, p. 767).

Data science involves many fields, and it should be taught that way. These include, but are not limited to, computer science, statistics, ethics, policy, humanities, and the social, environmental, and natural sciences. Data exploration, preparation, representation, and transformation have a broad reach, from mapping in geography to visualization in design studies. Science about data science has much history in science and technology studies, library and information science, and other fields. As explored earlier, the arts and humanities have contributions as well. Data science programs and their graduates ignore these areas at their peril.

Well-designed data science programs go beyond number crunching and button-pushing (Manyika et al. 2011). When done well, data science programs teach the basics of big data and show students how to marry data and methodological considerations to concepts and domain knowledge. Demand for data scientists handily outstrips supply, although much of this work is basic data handling that is (ironically) disproportionately susceptible to automation of the kind for which big data is known (Henke et al. 2016). Most data science programs fall short because they "demonstrate a focus on theoretical foundations and quantitative skills with relatively little preparation in domains outside of computer science, statistics, and mathematics" (Oliver & McNeil 2021, p. 12). These programs are missing courses in fields that regularly contribute to data science and disciplines that consider more significant issues like policy and ethics (more in Chapter 5).

Data science education is often tied to data science centers and programs in universities. The National Strategic Computing Initiative (NSCI), led by the NSF, focuses on solving software and hardware changes in cyberinfrastructure and funds training programs (Towns 2018). These include workforce-building efforts designed to increase the capability of the scientific researcher to develop or adopt advanced cyberinfrastructure, high-performance computing, and big data. The Moore–Sloan Data Science Environments (MSDSE) program offers a helpful overview of data science in research universities. The Gordon and Betty Moore Foundation and the Alfred P. Sloan Foundation joint project was designed to create favorable conditions for data-driven discovery at academic institutions (DeMasi et al. 2020). The program funded centers at three American universities, with US$12.5 million each to support data science environments for five years. The program found data science activities, curricula, and centers at around 80 percent of US institutions. The MSDSE program found that successful institutions tend to offer dedicated space, efforts around collaboration and interdisciplinarity, and community-building (Katz 2019).

Another thread in data science education is the idea of professionalizing the practice. Proponents for professionalization are groups who seek to formalize (and exert control over) data science, including the Association of Data Scientists, Data Science Council of America, Data Science Society, and Data Science Association. These organizations joined more established academic ones that embraced data science, including the Institute for Operations Research and the Management Sciences, and the American Statistical Association. These organizations have weighed in on data science credentials or other forms of professionalization. The basic argument for these activities is that other fields like medicine and law became professionalized in large part because there are severe harms to people who do not follow professional codes of practice (Walker 2015). While data science can be the source of harm, as explored in Chapter 5, extending the notion of harm to policy or other decisions flowing from poor data science may be a reach. There is pretty deep agreement in fields like law and medicine on what constitutes core knowledge and clear standards on ethical practices. The harms are also apparent, such as illness and death for medicine or incarceration for law.

In contrast, there is only emerging and preliminary agreement on what data science is as a field, let alone what should be core principles or how to quantify harm flowing from bad data science. Peer review, data sharing, and better replication are good ways to professionalize data science (Section 3.4 examines reproducible science). Poor data science in the private sector tends to be punished by negative market performance, policy interventions, or lawsuits (although there is room to improve these mechanisms, per Chapter 5 on policy issues). Walker's analysis may underplay other reasons for professionalization in other fields, especially the desire to raise barriers to entry to what are effectively guilds. However, the article recognizes how "self-regulation and setting performance standards improves the quality of services and protects the public notwithstanding secondary consequences of pay inflation or competition limits" (Walker 2015 p. 15). While data science can trace antecedents back decades, its modern form is still coalescing.

4.5 Focus: Smart Theories for Smart Cities?

Cities are among the most complex human-environment systems and the site of conversations around theory development in data science. Definitions vary, but smart cities are typically urban landscapes blanketed with sensors that track many aspects of the human and environmental landscape and feed into semiautonomous systems that can monitor and guide systems ranging from transportation and energy to schools and social services (Kandt & Batty 2021). Like many aspects of big data, smart city elements have existed for decades, such as sensors that monitor traffic, air quality, or hydrology. Indeed, technologies such as the telegraph, early census data, and telecommuting can help us trace smarter cities back decades and even centuries (Townsend 2013). While much of the focus is on aspects of smart city information and computer technology, the term "smart" also carries broader connotations of intelligent policy approaches to urban issues (Lazaroiu & Roscia 2012).

There is hope that smart cities can help increase underlying scientific knowledge and advance theory in urbanization and human-environment systems. Batty (2013) lays out a compelling case for the potential for a new science of cities that will use the volumes of data to increase our knowledge of how cities work. He and others (Batty et al. 2012) lay out a research program for smart cities (Table 4.1). The work is notable in how it seeks to balance the blind technological utopianism that marks so much rhetoric around smart cities versus the oft-times deep skepticism of smart city critics.

A pivotal axis for debate around the role of data science in urban science is the task of computationally modeling cities. The classic article, "A requiem for

Table 4.1 *Smart cities research goals and data science*

Research goal	Data science roles
Relate the infrastructure of smart cities to their operational functioning and planning through management, control, and optimization	Data science can recognize how information and communication technologies work into the fabric of cities and may be yoked to big data in ways that improve quality of life by controlling a range of services and operations in real-time
Explore the notion of the city as a laboratory for innovation	This idea has a couple of different facets. However, essential is the idea that big data is being used to make many systems more efficient, and there is room to take a more active role in manipulating and experimenting with how these data and systems interact with real-world features of cities
Provide portfolios of urban simulation that inform future designs	Urban simulation modeling has been around for decades, but big data is breathing new life into the enterprise
Develop technologies that ensure equity and fairness, and realize a better quality of city life	What role does big data play in balancing efficiency with equity? Much has been written on how big data may be increasing inequality. However, there is also cause for hope in its potential to democratize many aspects of society or aid governments and firms in recognizing equity issues
Develop technologies that ensure widespread participation	Data science, driven by computerization and networking, is double-edged because it offers expanded roles for citizen participation while also positing privacy and security concerns
Ensure and enhance mobility for urban populations	Mobility is the heart of urbanization, be it movements through space or connecting vitality. Accessibility to work, leisure, and other opportunities is key to whether big data helps or harms citizens

(Research goals from Batty et al. 2012).

large-scale models" argued that urban planners were overly invested in developing and using large mathematical and computer models to plan cities (Lee 1973). These models invariably focus on the wrong data and ask the wrong questions. Data scientists risk repeating these mistakes decades later with smart cities. The tools and data are better than previous versions, but they will need the entirety of Batty's science of cities and its cautious optimism to succeed.

What theories should data science apply to the study of smart cities? There is likely room for *middle-range theory* that examines cities via the lenses of science and technology studies, complexity science, and sustainability studies that embrace a nondeterministic view of how smart cities will evolve (Mora et al. 2020). This work is based primarily on management and technology literature and focuses on urban transitions in time and space. There is also room for the very abstract modeling conducted in physics and mathematics to be tied to practice on the ground and existing theory in urbanism, society, and the environment (O'Sullivan & Manson 2015). Data science runs the risk of ignoring a vast amount of existing research while advocating for a unified theory of urban living defined by an "integrated, quantitative, predictive, science-based understanding of the dynamics, growth and organization of cities" (Bettencourt & West 2010, p. 912). This work is exciting in its vision for a sharper focus on the new science of cities and moving beyond the fractured nature of urban studies. However, the chances are that it must incorporate the large existing body of urban scholarship to have the highest chance of success.

Urban analytics and smart cities are a site of promise and contestation around managing coupled human-environment systems. The promise lies in using computers to create faster feedback-driven decision-making. For example, there is great interest in *digital twins*, computational simulacra of actual systems such as traffic control, hydrological networks, and air quality as a function of traffic (Nochta et al. 2021). The idea here is that big data allow real-time monitoring of issues, such as snarled traffic or a flood, and data science will facilitate the digital twin to direct real-time responses, such as shifting traffic signaling or opening sluiceways. Smart decision-making could save hundreds of billions of dollars in unnecessary expense while increasing human well-being. Better decisions could also enhance urban environments and reduce the global ecological footprint of city dwellers through such actions as reducing the climate impacts of transportation or consumption.

Many of the generic pitfalls of data science apply to smart cities research and theory development. A case in point is how some smart city infrastructures and reporting systems succeed; other seemingly well-supported technological efforts fail in the face of workaday bureaucratic cultures and politics in ways that are hard to generalize (Kitchin 2014b). It can be challenging for a city with established governance and politics to wrest primary data from a range of bureaucratic fiefdoms. Personal biases and politics can get in the way of heeding these data, and the price of cutting-edge systems may be out of reach for many towns and cities (Sadowski 2021).

It is easy to imagine some flaws or issues quickly propagating in flood control or transportation planning domains even when data and computing work well. The danger with many real-time systems is that they usually imply real-time decision-making, which may mandate automated behaviors mediated by computers. For example, there is a good deal of interest in using big data to understand, predict, and eventually shape modal choices, the various kinds of transportation that people choose (Omrani 2015). As explored in Section 4.1 on theory in data science, many essential data science tools remain opaque to humans, and as Chapter 5 describes, algorithms control people's lives in ways that we are only beginning to understand. Managing cities with data science absent underlying processual theories and understanding may lead to unforeseen and undesirable outcomes.

Theory development can help guide research and policy when the rosy visions of a happy future meet messy reality. Normative concepts of smart cities are driven in part by a European Commission effort to define smart cities in terms of renewable energy, energy efficiency, and environmentally friendly forms of transportation (Lazaroiu & Roscia 2012). From this drive, specific characteristics and elements (or indicators) can be used to measure smart cities and guide policy. Chapter 5 delves into policy and these issues more generally, but many features of smart cities are heavily associated with a long history of gentrification (e.g., the attractiveness of natural conditions, innovative spirit, productivity, touristic attractiveness). Others, like individual safety, have been conflated with violence and racialized policing and are finding new expressions in data science. For example, the growing use of data science can be used to criminalize, control, surveil, and push out long-time urban residents (Jefferson 2020).

Similarly, the call to apply data science to urban challenges often ignores the power dynamics inherent to who is allowed to name these problems and their chosen solutions. The explicit appeal to financial benefits for smart cities assumes that resources flow to commercial firms that are creating technologies and guiding the choice of where they are applied (Mattern 2013). At a more fundamental level, data-driven approaches are often embedded in, or embody, instrumental rationality that can institutionalize negative social dynamics, including stigmatization and bias based on race, class, or gender (Hollands 2015). A theoretically informed under-standing of the city – drawing on the environmental, information, and social sciences, and the arts and humanities – should contextualize the decisions made with data. The smart city should be centered on the lifeworlds and lived experiences of the people and places that make up the urban patchwork (Travis & Holm 2016).

Smart cities also risk reifying existing physical and social stratification. Wealthier and politically well-connected neighborhoods are more likely to translate information into action, and so it is vital to work with local communities (Zegura et al. 2018). Part of their advantage comes from simply better technology, such as sensor placement and access to smartphones or personal cameras that act as an informal neighborhood watch. For example, there are many examples of where sensors and intelligent control systems are unevenly placed, being clustered in well-off regions. The smart city is heavily reliant on sensor infrastructures that measure a wide range of features in the urban environment.

Perhaps the most visible expression of the smart city, sensor deployment is a key equity concern. As new sensor technologies and Big Data interact with social processes, they have the potential to reproduce well-documented spatial injustices. Contrary to promises of providing new knowledge and data for cities, they can also create new gaps in understanding about specific urban populations that fall into the interstices of data collection – sensor deserts.   *(Robinson & Franklin 2021, p. 238)*

Alternate conceptualizations of smart cities may be needed to enact data science in the Global South. Current conceptualizations of the smart city often rely on many features of highly industrialized or postindustrial cities. Actual levels of data handling capacity and citizen participation in smart city initiatives can be markedly different than assumed (Smith et al. 2022). Attention must also be paid to their *data cultures*, specific historical and cultural circumstances of cities and periurban regions in any given area (Cinnamon 2022). Additionally, smart cities are usually couched in terms of Western and industrialized tropes that can perpetuate a colonial mindset (Ash et al. 2016). Datta (2015) looks at the role of big data in developing Dholera as the first Indian smart city. Prioritizing urbanization as primarily a business concern elides many realities on the ground, including a contested history of development, and ignores many of the local voices and interests who are the ostensible beneficiaries of the smart city. Section 5.2 examines digital divides between the Global North and South, and Section 5.4 explores what data science means for sustainable development in many places.

Ultimately, human-environment researchers may need a mix of theories to guide their scholarship on urban social and natural dimensions. There is a fundamental tension between the "high frequency data and the long durée of structural challenges facing cities" (Kandt & Batty 2021, p. 1) because much of what occurs in a city is highly path-dependent. Today's cities reflect decisions made decades and centuries ago. This tension results in some places building new cities from the ground up, weaving information infrastructure throughout. For example, South Korea is working with Cisco Systems and spending billions of dollars on building Songdo, a model city centered on energy conservation achieved through big data from millions of sensors embedded in roads and power grids to track and predict people's movements (Moses 2013). Coherent theories of urbanization should grant insight into cities past, present, and future, and inform the promising new science of cities.

4.6 Conclusion

Debates over the role of theory in data science reflect discussions that have existed for centuries. Many scholarly disciplines have some version of a debate that maps onto the dyads of deep versus shallow, narrow versus broad, universal versus particular, generalization versus specialization, or nomothetic versus idiographic (Rieppel 2010). Nomothetic approaches generalize or derive laws, often assuming that phenomena can be objectively measured and explained. In contrast, idiographic work is predisposed to seeing phenomena as contingent, unique, and deeply subjective. At the risk of being glib, the exact nature, tenor, or resolutions of debates about the nature of knowledge do not matter.

What does matter is that scholars and practitioners of every stripe should strive to see the bigger epistemological picture and understand the work of other researchers in other fields. Not doing so exposes scholarship to the danger of demonizing specific approaches or knowledge. Of course, every researcher must develop their view on doing their work, and some approaches will end up being better than others for any given topic. Nonetheless, the way forward is accepting multiple approaches (while guarding against the excesses) and trying to find productive shared spaces when developing theory for the data science of human-environment systems.

5

Policy Dilemmas

Data science holds extraordinary promise for better policymaking for humans and the environment while at the same time posing myriad harms. Data science helps draft solutions to many human-environment problems, including climate change, environmental degradation, and sustainable development. It also has many troubling aspects. It is reshaping many human activities in ways that can increase their negative environmental impacts. It is remaking many facets of society, including injuring privacy, increasing surveillance, and negatively affecting people's lives. In response to these actual and potential issues, there is growing interest in developing scientific, legal, and policy mechanisms to protect sensitive environmental areas, expand open science, and address concerns around privacy, choice, and other social impacts of big data. Many of the policy dilemmas and solutions around data science are brought into sharp contrast by its application to achieving sustainable development goals worldwide.

5.1 Data Science Dilemmas

Data science boosters point to how big data and advanced analytics are a boon for human and environmental systems. Less examined are some of the dilemmas posed by this work. There is mounting evidence that data-driven systems can reify and create many forms of discrimination, although there is also hope for using data to combat inequity. Data science can threaten to dispossess people of data, gathering and commodifying information in ways that harm people. Surveillance is another area where many actors engage in problematic behaviors, even as some forms of surveillance data are increasingly integral to human-environment research. Finally, many people and organizations wrestle with challenges around privacy and consent in a data science context.

5.1.1 Data Science and Discrimination

Data science has a challenging relationship with discrimination, fighting it in some ways and propagating it in others. Data science methods that draw on discriminatory data, such as those steeped in racism or sexism, can readily create models that perpetuate discrimination in many social and legal contexts. A key challenge for data science is to eliminate discrimination in data, methods, and theory to create better research and deal with a significant social ill. Data science can go a step further and seek to root out such discrimination in many kinds of human-environment dynamics.

Big data have big biases. The weak link in many data science analyses is the choice of the training set. Formulating appropriate training features for machine learning involves many choices. Who is included in data? Using an overly narrow set of input variables or the choice of proxy variables, such as race, language, or class, has been shown to create unintentionally discriminatory machine models (Žliobaitė & Custers 2016). Even large and reasonably complete data sets cannot fix how machine learning and related approaches are outstanding at modeling relationships in data. The problem is that, along the way, these methods reify existing forms of discrimination encoded in policy and legal systems (Brayne 2017).

Legal frameworks around discrimination in most places around the globe are narrow and not especially responsive to data science challenges. Legal protections around discrimination in many countries are typically concerned with the idea of intent instead of outcome, which makes proving or responding to discrimination complicated (Kroll et al. 2016). Discrimination can be direct or indirect; for example, discrimination based on race occurs when a system or rule treats a person differently due to their race, while indirect discrimination impacts a person of a given race incidentally. It is a subtle but necessary distinction because many legal systems prohibit direct discrimination while giving much more latitude to indirect discrimination, particularly linking intent to discriminatory acts (Barocas & Selbst 2016). Discrimination also has many forms that remain understudied. For example, most legal discussions of racism tend to be blind to, or at least minimize, the structural racism inherent to most forms of social and economic organization (Virdee 2019).

There are emerging policy solutions to combat discrimination in data science. Researchers and organizations can audit machine learning or data mining outcomes (Selbst 2021). Related are efforts to ensure consistent human oversight at all stages of using data science approaches. This supervision is particularly important when decisions are made, ideally with the involvement of outsiders or trusted third parties to assess outcomes. There is also value in assessing workflows by using, but not sharing or releasing, potentially sensitive information about race, gender, and other personal characteristics (Veale & Binns 2017).

There is also much interest in legislation for personal data protection or offering tighter safeguards against discriminatory behavior that results from data science approaches. For example, the European Union is promulgating directives about explainable and transparent data science models (Voosen 2017). These discussions seem far from some kinds of research. However, many forms of human-environment inquiry can lead to potentially discriminatory outcomes if they use biased data or encode problematic practices or policies. It is incumbent upon scholars to engage with these issues in their work.

Researchers are responsible for fighting discriminatory data science, especially since many standard legal and policy mechanisms are behind the times. There is a broad array of good research on how data science leads to discrimination, but it is often abstract. It may not offer concrete advice to scientists dealing with the biases of historical data, poor or underexamined training classifications, or unrepresentative data sets (d'Alessandro et al. 2017). There is a need for more research on computational means for dealing with discrimination. Solutions include building less biased data sets, creating more readily interpretable algorithms, and sharing features of data sets to make them transparent (Kroll et al. 2016). Data protection legislation, such as the European General Data Protection Regulation (GDPR) of 2018, takes positive steps to minimize data gathering and retention or limit access to potentially sensitive personal data. Ironically, while these actions offer many benefits for the right reasons, they may make it harder to acquire and use the sensitive data necessary to investigate and prove discrimination (Favaretto et al. 2019). The evolving policy landscape will need to account for these complexities.

There is great potential for data science approaches to reduce or even actively combat discrimination. One great hope for big data is that it could remove human bias from many decision-making processes. While there is a danger of reifying existing discriminatory practices, a well-designed and transparent data science system could reduce the influence of human fallibility (Abebe et al. 2020). Models could do more to help root out discriminatory practices, finding subtle (and not-so-subtle) instances of people affected by existing systems. For example, formalizing and measuring a sometimes vague concept like discrimination can make it more visible and treatable. "It requires various forms of discretionary work to translate high-level objectives or strategic goals into tractable problems, necessitating, among other things, the identification of appropriate target variables and proxies. While these choices are rarely self-evident, normative assessments of data science projects often take them for granted, even though different translations can raise profoundly different ethical concerns" (Passi & Barocas 2019, p. 39). While quantitative measures only tell part of the story of discrimination, they can have value when forcing people to be explicit about linking measures to outcomes that can be assessed over time. Further, specifying what is going on in an algorithm can highlight a system's underlying social and environmental characteristics. Algorithms usually only do as well as the underlying understanding or specification of the system (Mayson 2018).

5.1.2 Data and Dispossession

Data science can dispossess people of their data. Data science is premised on a highly asymmetrical power relationship in which individuals are *dispossessed* of the data when firms take the economic value of these data (Thatcher et al. 2016). Accumulation by dispossession occurs when people's data are taken, gathered, and commodified. Data that were formerly private, such as the location of a health appointment or where people dine, become commodified. Big data proponents would argue that often these data have no value to an individual or that people are compensated by receiving free services, like email or social networking.

The counter-argument is that people give away something important, namely their control over how these data are used and cede control over experiences previously unknown to anyone not directly involved. Biopiracy is another example, where agriculturalists, predominantly in the Global South, have genetic information on their cultivars and agricultural knowledge taken by firms that subsequently use them to create modified patent-protected seeds (Moran & Lopez 2016). These companies may improve seeds and cropping practices at the cost of dispossession. The conversation around dispossession points to an ongoing need for better and more nuanced approaches to dealing with data and its value for various parties.

Dispossession is tied to information manipulation by social media firms and other organizations. The power to manipulate data and ideas poses profound ethical dilemmas for data scientists, who must make appropriate data use and protection choices (Selinger & Hartzog 2015). The extent to which this could be a problem was made clear by the case of Cambridge Analytica. This company used information from Facebook to compile records on users and target them with political ads. The breach, in this case, was that the company received permission to gather user data but also asked for permission to capture friends' data, which is a significant departure from ethical research standards (Schneble et al. 2018). Outside companies are not the only source of problems. Facebook purposely manipulated information in the name of experimentation to see whether it could shift political sentiments (Fiske & Hauser 2014). These are the issues we know about, so there are surely more.

Even when information remains private, there remain concerns that aggregated data on individuals can still be combined in ways that harm populations. In the extreme, social media platforms like Facebook or Twitter promote the creation of misinformation online by monetizing popular content, even when it is wrong or incendiary (Hao 2021). Such uses of data violate the autonomy of individuals when their data is used without their consent or knowledge but still has an effect on them (Francis & Francis 2014). Matters are complicated by how social media systems have complicated and obscure relationships with various other technologies related to location tracking, advertising, finance, search, and entertainment (van Schaik et al. 2018).

In the extreme, dispossession can mean a person loses control over their identity. People find themselves in a situation where their digital and physical identities are indistinguishable (Wright et al. 2008). For example, near-instant background checks are used to acquire financial credit, rent a car, or lease an apartment. These checks rely on multiple digital sources across social media and commercial databases to assess creditworthiness. In essence, a person's identity is only real insofar as it relates to an extensive sociotechnical apparatus beyond human reckoning (de Vries 2010).

People are increasingly interchangeable with their data, where data are used to construct profiles that are then used to quantify many features of a person and eventually affect that person. These profiles are often positioned as a boon, in that a store can tailor a shopping experience, or an advertiser can offer relevant ads. However, this personalization quickly moves into more complicated and fraught territory because it leads to people being denied services or discriminated against when the underlying algorithms are biased (Kitchin & Dodge 2011).

For all of the dangers of data defining people, being excluded from databases is also fraught. Scholars are interested in *data antisubordination*, or the right not to be overlooked in data. Antisubordination is important to people and groups who are not adequately represented in big data or are disadvantaged by policies and analyses implemented via data science (Lerman 2013). For example, credit reporting is used in the United States in new and sometimes harmful ways, such as setting insurance premiums or determining whether someone is employable. However, not having a credit report carries its own travails, making it hard to engage in elementary functions of society, like opening a bank account or applying for a job. Being invisible in data has advantages but increasingly comes with disadvantages as well. As human-environment research increasingly embraces data science approaches, it risks missing people who are not well represented in many data sets, perpetuating dispossession.

5.1.3 Surveillance

Many kinds of big human-environment data are gathered via private and public surveillance. Remotely sensed images or social media feeds are profoundly helpful in linking transportation to air pollution or tracking the effects on vegetation of urban growth. However, nearly continuous surveillance of people and many features of the environment can be problematic. While people and places have always been surveilled to some extent, the last few decades have seen explosive growth in private and public monitoring due to advances in computation and networking, the politics of fear, and pervasive personalization. Human-environment research benefits from many forms of surveillance, but these data come with increasingly fraught ethical dimensions.

Many people live in a surveillance society and have since the earliest human civilizations (Lyon 2001). Governments, in particular, have long tracked populations, conducting surveillance of individuals and specific groups and holding broad-based population censuses for decades and even centuries. To these have been added health surveys, information on migrants and people making border crossings, crime, and other features of the social and natural worlds. Guerry wrote *Essai sur la statistique morale de la France* (1833) as one of the first examples of modern surveying of a range of phenomena (e.g., crime, deaths, literacy) for French departments (second-level administrative units) in the 1820s and 1830s. The author organized these data as tables and maps, and analyzed them via multivariate comparisons, graphics, and thematic cartography. This work originated many of the tools that we use today to understand spatiotemporal processes (Friendly 2007). The gathering and use of social statistics is a form of surveillance.

Surveillance has moved well beyond the state's remit in the last century, and many organizations track our daily activities. Several driving forces are rewriting the underlying rules and assumptions about how government agencies, firms, and other organizations monitor individuals (Vardi 2022). As for data science in general, significant factors in the growth of surveillance are computing and networking (Section 1.4). For most of human history, most kinds of surveillance encountered physical limits. For example, governments have always employed an array of people to observe other people, including police, health workers, and census takers. However, this approach is labor-intensive and expensive, time-consuming, and therefore limited in scope.

Technological advances made surveillance easier by orders of magnitude. The 1940s and 1950s saw the advent of computers for keeping records and making connections among them, while networking connected computers to share information more readily and gather data from more places (Manson 2015). Miniaturization was also helpful because computers became faster and smaller.

Networks became common as telecommunications equipment became less expensive, lighter, and nearly ubiquitous in many parts of the world. Cell phones are an excellent expression of these three trends, as they are essentially small, networked, and nearly ubiquitous computers packed with sensors. They and other personal mobile devices are excellent surveillance tools.

The second driver of growth in surveillance is the rise of what has been termed the politics of fear. According to some basic measures of human well-being, including poverty or life span, the world is becoming a better place for a growing share of the global population. At the same time, however, we see greater promulgation and sharing of news about specific kinds of threats around terrorism and state-sanctioned violence. These events and systems help facilitate a *politics of fear*, the purposeful stoking of fear or worry in a population to help achieve political goals (Altheide 2006). One of these goals is to broaden commercial and governmental capacity for and acceptance of mass surveillance, such as cameras placed in private and public spaces in the name of safety.

There was an expansion in governmental use of surveillance that went hand in hand with the expansion of the technological ability to perform surveillance, ranging from tracking people via their electronic trails to increased use of cameras in public spaces. The United Kingdom's Investigatory Powers Act (2016) and the United States' PATRIOT Act (2001), for example, increased the power of the government to track individuals through a range of approaches, including their phone records and online activities, with less judicial oversight than was previously standard. At the same time, a series of whistleblowers, including the high-profile case of Edward Snowden, broke the news that several governments were illegally collecting a broad array of data and metadata on citizens (Waranch 2017). These instances are just a few ways that data are collected in vast quantities and repurposed for many reasons, including surveillance.

The third driver of surveillance is pervasive personalization. Surveillance has traditionally involved tracking a relatively small number of people over space and through time or greater numbers of people at fixed locations such as border crossings or via large-scale surveys like the census. Surveillance has become pervasive and personalized. In many parts of the globe, especially in cities, a single individual's face (or vehicle license plate) is recorded dozens of times per day. These images are fed into databases and are automatically matched against existing data. Almost every online interaction is recorded by one or more parties, from internet service providers to websites to servers hosting companies to government agencies (Section 2.2.6 introduced some of these issues). The trend toward pervasive personalization is best exemplified by our relationship with the cell phone, an almost perfect surveillance device. It is usually linked to a single person and possesses a microphone, camera, and location tracker tied to persistent network access, and is monitored by many organizations.

Surveillance data are integral to human-environment research. Scholars benefit from a global drive to quantify almost everything that transcends specific uses and focuses instead on developing indexical information on the environment and people (Chapter 2). Big data have elevated the potential opportunities and challenges of surveillance. For example, lidar and remotely sensed data can inadvertently expose vulnerable animal populations to poaching or culturally significant artifacts to vandalism (Arts et al. 2015). At the same time, scholars argue that data science approaches like combining remotely sensed imagery with semantic mining of news feeds may offer an early warning system to protect unique locations such as world heritage sites or environmentally sensitive areas (Levin et al. 2019). Surveillance offers challenges and opportunities to researchers.

Researchers must make ethical decisions about their data and research. State surveillance often has very few limits, and even these relatively modest rules are broken with disturbing regularity. China is getting much attention for integrating disparate data platforms into a surveillance infrastructure designed to monitor and predict the trustworthiness of people and organizations (Liang et al. 2018). However, it is just one of many nations with robust surveillance programs. Many private-sector firms collect, aggregate, and analyze data to monitor people and places. Commercial surveillance has comparatively little oversight, and in any case, most users from which data are collected have unknowingly signed away many rights. The Cambridge Analytica scandal and related incidents have shown that even these meager limits are ignored when convenient (Schneble et al. 2018). Firms trade data to enrich their customer profiles and draw on public data, which they either collect at the source or purchase outright from local, regional, or national governments (Kitchin 2014a). While relatively little human-environment research is conducted with secret surveillance data, there is some, and it is growing. Declassified military satellite imagery, restricted census data, protected social media information, or bespoke commercial data are examples of information that finds its way into valuable human-environment research.

5.1.4 Privacy and Consent

Privacy and consent are contested entities in the era of data science. Privacy is often seen primarily in terms of legal and technical controls over the flow and disclosure of information. It can be socially oriented, where interactions among people texture how much is shared, when, why, and with whom. Privacy can be hard to define but involves day-to-day tradeoffs that balance risk, hassle, and payoffs (Crabtree et al. 2017). Widely shared privacy principles exist, but many of the features of big data make them hard to enact. Scholars and research institutions face additional hurdles around regulatory oversight and changing norms around privacy and consent. Finally, while most conversation (and this section) focuses on human privacy, note that a growing body of work argues for the rights of nonhumans, including the rights of animals, to privacy (Pepper 2020).

Privacy can be framed in terms of harm people experience at various stages of the data science chain. What are the negative impacts of losing privacy (Solove 2008)? These start with the act of information collection, particularly surveillance, and the role of technology in surveillance. The next step is information processing and the harms that may come from data aggregation or when data are repurposed in ways not anticipated or known when they were collected. Harms that accrue during the data dissemination stage can include breaching confidentiality or disclosing potentially embarrassing information about a person. There is the concept of intrusion into one's peace of mind or personal affairs. Even if a person believes their data are safe from accidental disclosure, there is always the subtle and insidious knowledge that eventually those data may be shared or stolen. Many harms flow from a loss of privacy.

Data science challenges existing standards around privacy. The Organisation for Economic Co-operation and Development (OECD 2013) developed a set of privacy principles in 1980 (clarified but essentially unchanged in 2013) that serve as a foundation for many legal and policy frameworks. Big data and data science pose problems for these principles by their nature. The pace of computationally driven privacy threats has outstripped the remit of many privacy laws and expectations. These principles relate to data quality, purpose specification, openness, individual participation, accountability, security safeguards, and use limitation. It is instructive to consider these principles (paraphrased here from the 2013 OECD document) in light of data science.

Data Quality. *Personal data should be relevant to the purposes for which they are to be used and, to the extent necessary for those purposes, should be accurate, complete, and kept up-to-date.*

As discussed in Section 2.3, among the draws of big data are their volume and potential to be combined with other data to, in essence, create new data or be repurposed in ways never intended. As a result, the way that scholars work with these data is almost inherently at odds with the data-quality principle, mainly because all data have bias, error, and uncertainty. As discussed later, the idea that data should be relevant to a specific purpose is antithetical to how data science is more effective when it combines multiple data sources.

Purpose Specification. *The purposes for which personal data are collected should be specified no later than at the time of data collection; subsequent use should be limited to fulfilling those or compatible purposes.*

Big data almost inherently contravene the purpose specification principle because they are often used in new ways and to novel ends. For example, one of the most promising data sources for human-environment research, cell phones, is also the most problematic in how it links specific people to their location and behavior. There are many potential ways cell phones and their attendant data can pose privacy concerns (Christin et al. 2011). Time and location may be used to infer a person's health, religious preferences, or personal behavior (Gonzalez et al. 2008). As explored in Section 2.2.5, mobile devices also open doors to other information. For example, inertial sensors can be used with external video feeds to develop unique gait signatures to identify individuals. Similarly, biometric data such as heart rate can assess health, especially when combined with sleep or physical activity information.

Purpose specification is demanding with ostensibly public data. When are personal posts, blogs, tweets, or other social media considered public? What are expectations around readily collected remotely sensed imagery? There have been several high-profile cases where researchers have used ostensibly public data that users later contended were never meant to be shared (boyd & Crawford 2012). For example, there is a gap between citizen scientists purposely collecting data for a specific use, like tracking outdoor recreation, and collecting the same data without consent (Connors et al. 2012). Publicly available data may not be considered private or even protected when the information is identifiable or individuals have no reasonable expectation of privacy. The tricky part for data science is how people expect or believe their social media posts are private. Social networks usually have access controls limiting data shared with the broader public, but these restrictions are typically inconsistent with standard research protocols (Nadon et al. 2018).

Openness. *There should be a general policy of openness about developments, practices, and policies applied to data, with attendant ways to establish their existence, nature, and purposes.*

Openness is difficult to enact in many data science contexts. Private firms are interested in preventing competitors from learning what they do with data. These data are the source of their profit. In contrast, while some public agencies have an ethos of openness, others often have no duty to openness around their data-handling procedures or policies. Section 5.3 delves into open data issues.

Individual Participation. *A person should be able to confirm their data exists and obtain them within a reasonable time and cost in a readily intelligible form. If not possible, the person must be able to challenge the denial and, if such a challenge is successful, to have the data erased or changed.*

Individual participation is too often a function of the person and situation. Even the most technologically savvy and well-resourced individuals can face challenges in devoting time, expertise, and effort to learning about their data despite ongoing concern over their use (Leszczynski 2015). Social media firms and others, like credit reporting agencies and potential employers, are getting better about sharing which data they have on individuals, but almost exclusively in response to legislation or immense public pressure.

No individual has the ability or power to check or control all the information held about him or herself, yet this information has the power to remove financial credit, curtail the movements of an individual – indeed, to ruin lives. The levels of intrusion into personal privacy are likely to grow – as are the risks of data breaches, accidental losses and deliberate thefts which have become simpler due to technological innovations in information storage.

(Elahi 2009, p. 117)

Accountability. *A data controller should be accountable for complying with measures that give effect to the other principles.*

Like the other principles, accountability has a very mixed record. The nature of redress for lapses in accountability is contingent on the jurisdiction where the individual and data entity reside. Even with clear legal standards around data breach disclosures or other issues, accountability is low or incomplete, even in jurisdictions like the European Union with solid regulations (Nieuwesteeg & Faure 2018). Matters are complicated by how data are free to move across spatial, temporal, and political contexts, navigating a complex mishmash of rules and norms. There is only partial agreement among standards, including the OECD guidelines on the Protection of Privacy and Transborder Flows of Personal Data, the US Consumer Privacy Bill of Rights, or various efforts in the European Union and other regions.

Security Safeguards. *Reasonable security safeguards should protect data against such risks as loss or unauthorized access, destruction, use, modification, or disclosure.*

Ensuring data security is an enormous task that can impose research costs when successful. It is possible to use big data to impute sensitive personal information from various data sources. Even relatively simple combinations of date of birth, gender, and post code identify many people in many places in the United States, although these estimates may be overly protective given the issues with source data for these analyses (Krzyzanowski & Manson 2022). There are many technologies designed to preserve privacy based on cryptographic techniques implemented through software and hardware for all stages of the data science cycle, from security and privacy during data acquisition through to deidentification in analysis and methods designed to work on encrypted data (Jain et al. 2016). Standard social science approaches include using pseudonyms, batching or recoding data to generalize the information, and adding noise or perturbations to the data, with potential downstream impacts on data quality unless handled well.

This security can limit research. For example, ensuring that relatively rare subgroups in data are not inadvertently identified (e.g., by ethnicity or race) can make these people invisible in any subsequent analysis. Other approaches, like differential privacy, can go quite far (or too far) in purposely and systematically adding errors in the name of security. The 2020 US Census applied differential privacy to much of its data even when it was not clear that these measures were necessary (Ruggles & van Riper 2022). Moreover, differential privacy fundamentally degraded these data's utility for human-environment research, including introducing disproportionately discrepancies in many basic statistics for rural and nonwhite populations at the levels of analysis used by many scholars and policymakers (Mueller & Santos-Lozada 2022).

Use Limitation. *Personal data should not be disclosed, made available, or otherwise used except with the subject's consent or by legal authority.*

Use limitation is often tied to issues of consent in academic research. Regulatory agencies and academic review bodies struggle to remain current. In the United States, most human-subjects research is moderated by *institutional review boards* (IRBs), administrative bodies designed to protect the rights and well-being of human research subjects who participate in research activities. Staff members of IRBs know that data science is growing in popularity and remain optimistic about dealing with attendant challenges. However, a consensus is still emerging, and there is an ongoing need to increase staff knowledge and technical capability (Vitak et al. 2017).

Current research oversight relies almost exclusively on entrenched top-down regulatory frameworks or the threat of legal action. These regimes can fail to capture the nuances of responsible research or may encourage mindless adherence to rules that satisfy rules while ignoring intent. In simple terms, few institutions have rules explicitly designed to account for the vagaries of big data research: "most universities are simply not prepared to perform ethical evaluations of research proposals that make use of vast amounts of data collected from social media, secondary apps and the Internet" (Schneble et al. 2018, p. 1).

Many scholarly trappings to protect subjects are challenging to apply to data science. Tools such as informed consent or analog methods for ensuring privacy (e.g., only having paper copies of respondents' personal information) are not sufficient for the task or do not apply to many data science contexts. These are not arguments against these protections. Ioannidis (2013) notes that any data relevant to research should have informed consent protection, even in cases where data science approaches are being used to reanalyze or extend the analysis of data collected for another purpose. Unfortunately, the onus is on the research community, not companies, to rework informed consent processes to allow for later and unanticipated uses or develop approaches that preserve confidentiality during all stages of the research process (King 2011).

Institutions and researchers are taking steps to make better rules around consent. Researchers share widespread agreement on some elements, including removing subjects from data on request or sharing results with participants. There is disagreement about others, including whether it is reasonable to ignore the terms of service of online platforms or deceive participants when deemed too inconvenient for the study to do otherwise (Vitak et al. 2016). One potential alternative to rules around consent is to develop an impact model of ethics. This approach encourages researchers and technology developers to take a more active role in designing privacy and other safeguards into their work, ones that better take into account the actual positive and negative impacts of their work (Markham 2018). Scholars take privacy seriously, and hopefully, their institutions will catch up to the fast pace of data generation and use.

5.2 Digital Divides

The *digital divide* is when different people or places have different levels of access to digital affordances. The concept has a long and complicated history. It is a useful framing device, although it can sometimes focus too much on simple access measures and not enough on more complex definitions of accessibility. People may have good internet connections but lack the financial, social, or technical wherewithal to leverage this internet access. Moreover, many of these divides are perpetuated by the very technologies that define them (Crutcher & Zook 2009). Technology has opened many new possibilities for engaging with the world. However, there can be significant disparities in who has the technology, skills, and time to use technology in general and data science in particular. The digital divide plays out at multiple scales.

Significant digital divides exist at the global scale. Wealthy nations often have faster and less expensive internet services, more comprehensive or widespread access to computers, and more resources devoted to creating human-environment data than lower-income countries. For example, the cost of broadband as a function of income varies geographically (Figure 5.1). Up to a quarter of the world's population does not have regular access to electric power. The International Telecommunication Union (ITU) is the United Nations agency for information and communication technologies. It is found that the percentage of households with internet access is about 58 percent globally, with a high of 82 percent in Europe to

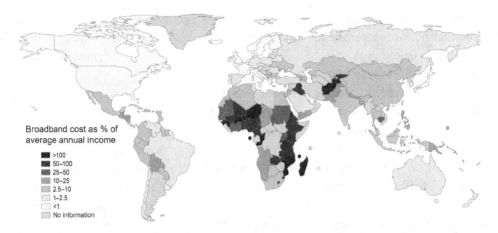

Figure 5.1 Cost of broadband subscription as a percentage of average yearly income (Graham et al. 2015). Reprinted with permission of John Wiley & Sons; CC-BY 4.0 (creativecommons.org/licenses/by/4.0/). Modified to render half of map in grayscale.

a low of 22 percent in Africa (International Telecommunication Union [ITU] 2019). Although 95 percent of the world's population lives within the reach of a mobile internet network, 37 percent of the global population does not have access. That is 2.9 billion people!

There are many differences within these global gaps. There are also notable divides in the use of technology that may fall along socioeconomic and demographic fault lines. Internet use in urban areas is twice that of rural areas. There are also age gaps, where 71 percent of the world's population aged 15–24 uses the Internet versus 57 percent of the rest, and gender divides, where the proportions are 62 percent of men versus 57 percent of women (International Telecommunication Union [ITU] 2021, p. iii). Gender gaps in the lowest income countries are even wider, with only 19 percent of women online versus 31 percent of men. Over time, discussions of the digital divide have shifted from focusing on access to basic technology, such as electricity, internet service, and computing, to more complicated divides defined by digital literacy and skills (Kelley 2018).

The digital divide affects human-environment data. Who owns smartphones or other devices that leave a digital footprint? Who has the time and digital savvy to volunteer human-environment data via crowdsourcing? Citizens more often produce these data from wealthy, tech-savvy urban neighborhoods than from poor, rural places (Section 2.3.4). The digital divide manifests as significant discrepancies in the quantity and quality of data production and access for different parts of the world. While there is much variation across countries and research domains, data generally have lower spatial and attribute resolution in lower-income countries due to fewer resources for data collection. These imbalances lead to gaps in coverage, accuracy, and timeliness for different regions (McCarthy 2016). As a result of this complexity, how the digital divide plays out must be seen as having many different dimensions or facets.

The potential for human-environment data science is contingent on digital divides. Greater access to technology like cell phones may offer the possibility for more widespread participation in data creation and use. It does not guarantee that actual participation will be equitable or evenly distributed without supportive social, economic, and political institutions (Warf 2013). There are also divides among groups in terms of finances, infrastructure, and expertise to develop data science workflows, which will affect the kinds of collaborations scholars can develop with these groups (McCarthy 2016).

Collaborations designed to reduce some digital divides can threaten to increase others. For example, open data environments encourage data sharing but can "lead to a broadening of the gaps between research institutions in rich and poor countries. Institutions from developing countries could become mere data providers, while wealthy institutions would apply state-of-the-art equipment and well-paid researchers to generate the new scientific breakthroughs" (Maeda & Torres 2012, p. 411). Reducing actual and potential inequalities requires intentional institution-building designed to go beyond equal access to data and embrace the development of tools and expertise to take advantage of these data.

Moreover, having and using data is not especially helpful if people cannot act on these data. Given that many data sets and methods are increasingly available via shared infrastructure, there is much potential to reduce digital divides, although change is slow for many regions. Additionally, the top-down nature of technology development in public and private sectors often does not adequately recognize the role of bottom-up communities of practitioners and users in developing a vision for bridging the divide (Gurstein 2003). These and many other issues come to the fore when considering the data science of human-environment systems applied to sustainable development (Section 5.4).

5.3 Data Open and Closed

The degree to which data are shared affects the quality of data science research. As discussed in Chapter 2, human-environment data comes from a mix of public and private sources, which practice varying degrees of openness. Apart from issues around privacy and commercial advantage, there are clear arguments for opening up data. These include supporting reproducible science, making good use of existing data and research infrastructure, and ensuring that people on either side of digital divides have the opportunity to use data. Much of the conversation around open data centers on publicly funded data and research. Significantly advancing big governmental data will require integrating it into national priorities and developing collaborations among nations for many kinds of data (Kim et al. 2014). Key issues for public open data include promulgating standards and the relationships among data, national well-being, and informational sovereignty.

A key driver for open data is the widespread policy-driven shift toward open science in academic communities. International organizations such as the Open Geospatial Consortium (OGC), International Society of Digital Earth (ISDE), Global Land Project (GLP), and the Intergovernmental Platform on Biodiversity and Ecosystem Services (IPBES) join public and private agencies such as the United Kingdom's Wellcome Trust and the US National Science Foundation (NSF) in promulgating open science and open data standards. Government data can be considered open when it fulfills the *Sebastopol Principles*, named for a 2007 workshop in Sebastopol, California, where data experts and advocates described the necessary conditions for truly open data (Charalabidis et al. 2018). These describe data that are complete, primary when possible, timely, accessible, machine-processable, nondiscriminatory, nonproprietary, and license-free.

Meeting these open data principles can be difficult. Many commonly accepted measures of data openness are inadequate under the Sebastopol Principles. Much data is aggregated or in the form of overviews, and the raw or underlying data are not readily available (Wang & Shepherd 2020). More broadly, given the general paucity of standard intellectual property or data-use agreements, there is a need to develop generic policies that ensure that data sets are not burdened with agreements that make sharing unnecessarily challenging. Some researchers call on nations to reassess existing rules and reconfigure those that impair the ability of academic researchers to collect, share, and publish data that other entities do as a matter of course (King 2011).

Another way to assess data openness is how systems support effective data use. Data providers need to look beyond making information available via the Internet and address underlying digital divides (Section 5.2). Instead, several conditions must be met for people to use open data effectively. Data providers are placing more effort on developing "useful and usable applications for those for whom such applications might be of immediate benefit. What is needed is both access (bridging the [digital divide]) but also the means for effectively using technology to respond to real crises in health care, education, economic development, and resource degradation" (Gurstein 2003, p. 13). Making data and applications open and useable requires systems that can offer critical elements (Table 5.1).

Table 5.1 *Characteristics of effective open data use*

Content and formatting	Are data formatted in ways that make them easier to use? Much of the growth in big spatial data, for example, is driven by the adoption of standards like GeoJSON and well-known text (WKT). These standards enable straightforward encoding and take care of underlying issues such as defining coordinate systems
Internet access, hardware, and software	Internet access as a function of affordability, sufficient speed and quantity for larger data sets, and accessibility (including where connectivity is politically or culturally restricted). Beyond raw access is the need for computing resources that are sufficient to process, visualize, and analyze data
Computing and software skills	Access to data is almost meaningless without the ability to use them. Data science is defined by complicated manipulation, analysis, and visualization that requires training and skill beyond the reach of many researchers, let alone nonexperts. These issues suggest the need to develop easier-to-use tools
Interpretation and sense-making	Do users have the requisite knowledge and skills to use the data? Sense-making interprets new data in ways that combine it with personal and local knowledge. These processes require enough underlying knowledge or familiarity with the data and its overarching domain
Advocacy and governance	Data absent individual and community resources are less valuable than data that can leverage advocacy and broader tools of good governance, including the necessary legal, political, cultural, and socioeconomic precursors to translate data into action

(Distilled from Gurstein 2003).

5.3.1 Government and Open Data

Governments around the globe are increasing support for open data with some exceptions. Reasons for openness vary, but they tend to fall into three categories (McDermott 2010). First, governments seek to be more transparent and accountable by promoting open access to information. Second, such open data can encourage innovation and improve efficiency in many parts of society, including the public and private sectors. Third, open access and innovation can spur engagement and participation in civic life. Many countries embrace these rationales to varying degrees. The European Union, United States, United Kingdom, Japan, Singapore, and South Korea are a few among many nations that have launched high-profile web portals or other ways to disseminate significant data holdings (Kim et al. 2014). Other countries, such as Australia, have developed a broader data strategy that subsumes big data to identify data, especially analytics, central to government services and data-as-infrastructure (Whiley 2018).

Governments are investing in open data infrastructure. For example, the United States launched the Big Data Research and Development Initiative in 2012. The effort involved dozens of national agencies committing over US$200 million for big data projects across science, health, defense, energy, and Earth systems science (Kalil 2012). This initiative was followed by the creation of the National Strategic Computing Initiative (NSCI). The NSCI is meant to spur innovation across the entire federal government, including offering investments in research and workforce development in collaboration with academia and industry (Towns 2018). The NSF and other organizations, for example, fund the Global Environment for Network Innovations (GENI). This network offers researchers a large-scale experiment infrastructure, a virtual laboratory. It offers distributed systems supporting data science, including virtual machines and network components that can be manipulated and reprogrammed to develop and test new forms of infrastructure (McGeer et al. 2016).

Open data require the availability of shared instrumentation and measurement processes that allow researchers to monitor their experiments. For example, the Extreme Science and Engineering Discovery Environment (XSEDE) and its attendant research computing resources and researcher support are designed as a nation-spanning set of supercomputing and other platforms (Towns et al. 2014). The argument for XSEDE and similar platforms is that federal funding agencies invest in computing support, but only a fraction is openly coordinated and shared in ways that make these investments efficient. Federal agencies like the NSF makes enormous "investments in resources and services that are restricted to use by particular projects and the associated scientists, limiting the options for leveraging these investments by an integrated national ecosystem" (Towns 2018, p. 43). Too many scholars receive funding for bespoke computing clusters that are underutilized, difficult to maintain, and start lurching toward obsolescence the day that they are created. Centralized support offers to make research computing more efficient and effective.

In addition to wanting to support research, the public sector invests research dollars into data science designed to protect national interests. Projects increasingly seek to link data across sectors to address future human-environment challenges. The United Kingdom's Horizon Scanning Programme (HSP), for example, hosts the Foresight International Dimensions of Climate Change Project. This initiative seeks to use big data from multiple sources to examine climate change impacts on food, water, and socioeconomic stability. Not surprisingly, these centers often focus primarily on climate change, as with the case of HSP, but can also consider the threat of terrorism and disease, as seen with the Risk Assessment and Horizon Scanning (RAHS) program in Singapore or the

Netherlands Horizon Scan Project (Habegger 2010). Many of these focal areas are entwined with coupled human-environment systems and are amenable to data science analysis.

5.3.2 Open Data Challenges and Informational Sovereignty

Open data are not without challenges around trust in the government. There is concern on the part of many open data advocates that the data are being used in ways counter to the goals of open data access. The Open Government Data policy in the United Kingdom is often held up as a prime example of doing open data well. It mandates that data collected by public agencies be made freely available for reuse, including those helpful to human-environment research, including weather, topography, and land use. Subsequent analysis demonstrates that these data are used by corporate and financial interests to further their own goals, often with the government's blessing (Bates 2014).

These situations may feed the suspicion that open data is a cat's-paw for efforts that could become antithetical to the open data ethos. More simply, citizens may be naturally skeptical when asked to provide data on themselves that could later be used against them. For example, the US Census has always been political. However, it became deeply so around the issue of whether and how to record citizenship status when the federal executive branch was making immigration a divisive issue (Levitt 2019). In the United Kingdom, people have grown skeptical of the government mandating that all local spending be made transparent and then using these data to curtail public spending or dismantle government programs in service of a larger neoliberal movement (Bates 2014).

Data are increasingly seen as central to *informational sovereignty,* the ability of nation-states to use data and analysis to manage socioeconomic and natural resources to serve their interests. Machlup (1962) was among the first to write about the knowledge economy, seeing how information steadily supplanted physical capital or resource extraction as a vital economic sector. Many governments restrict data dissemination because they want to sell the information or safeguard it for their own purposes. For example, the United States has seen many attempts to prohibit government agencies from publishing the weather and climate data that they gather. Instead, they would be made to give these data to firms that then repackage them for sale to the very public that funds data collection (Serra et al. 2018).

Additionally, there are legitimate concerns about data being degraded or made to disappear. For example, there is a push to rely on private firms for satellite data essential to weather forecasting, natural resource management, hazard response, and other critical public functions (Section 2.4). While the private sector offers innovation and efficiency, this move risks data quality and reliability, long-term data preservation, and data sharing when commercial interests control creation and distribution (Cirac-Claveras 2018).

Informational sovereignty is part of a broader move toward nations being *informational states* defined by their possession of, and policies about, information as a vital national resource. This movement traces back to the 1970s and the growth of such information technology as computing, web surveillance, personalization, and telecommunications (Braman 2006). The story of the joint Franco–German Symphonie project to develop communication satellites is instructive (Al-Ekabi 2015). The European governments planned to use their own launch vehicles but were stymied by technical failures in the early 1970s. They turned to the United States to launch the satellites. However, they were refused because the United States perceived it as a threat to their own burgeoning business in satellite communication. After the pair approached the Soviet Union for help, the United States relented and agreed to put Symphonie in orbit with the condition that it would only be used for experimental and not commercial purposes. These and similar events convinced the European countries of the need for domestic capacity in Earth observation, navigation, and communication. Remotely sensed imagery platform SPOT, the Galileo global navigation system, and a range of communication constellations owe their existence in part to the desire of European governments to control their critical data infrastructures. The decades following have seen similar moves by other countries, including Nigeria, India, Canada, and Japan.

Data science is central to informational sovereignty. Data themselves are valuable insofar as they enable other forms of power, such as programs that make economic trade or resource use more efficient. People use the phrase "data is the new oil" to describe the importance of big data (a cursory web search for this phrase in spring 2022 garnered hundreds of thousands of results). However, this saying misses the mark because oil is a physical good that can be produced in many places, and once a barrel is used, it is gone. Data differ from oil in how they are sometimes only produced in one place, as when a firm gathers data or creates new technology that is a trade secret. Data can be used repeatedly without destruction in some cases while losing value in others. An extension of informational sovereignty is cyber warfare or other forms of digital interference with a nation's capacity to use big data (Kaplan 2016). The capacity to create, use, and protect data is increasingly vital to national interests.

The legal and policy frameworks for data lag behind many technological and practical imperatives of data. Nations are regularly at loggerheads over how to handle basic privacy and data protection features when data are traded as a resource. It is conceivable that data-oriented agreements could be crafted in the mold of trade agreements that guide the flow of goods and services worldwide. Organizations, including the G20 and OECD, are taking tentative steps (Slaughter & McCormick 2021). In the meantime, nations have enacted comprehensive legislation to protect personal information, such as China's Personal Information Protection Law or the European Union's GDPR. These measures enact obligations in both private and public organizations. The United States has a mix of privacy laws but has not enacted similarly comprehensive federal privacy policies (Determann et al. 2021). Of course, these regulations often have exceptions for national security needs and have provisions for firms to house data in-country to make it easier to control and access.

Alongside data-centric legislation, many nations are making strategic investments in advanced computing in the name of national information sovereignty. Some countries and regions are making concerted national efforts to develop next-generation computing research projects, including China, the European Union, Japan, and the United States (Reed & Dongarra 2015). For example, China was among the first and the most successful in having international companies locate data centers domestically, allowing it to monitor data. Powering this system is machine learning and artificial intelligence, an area where China seeks ascendence for commercial reasons and in the interest of security and control, per Section 5.1.3 (Andersen 2020). These and other efforts are intentional and largely effective technonationalist visions that recognize the value of controlling data and associated technologies.

Weather insurance is a prime example of big data driving informational sovereignty and economic power. This sprawling and complicated market is composed of insurance and financial derivatives. Companies make bets in the billions of dollars on future weather events and how they affect energy or agricultural production (Bates 2017). Access to high-quality data has always been essential to these markets, but big data and data science methods are increasingly important. There are differences in access to national meteorological data that have long vexed the industry. For example, the United Kingdom is moving to more open data access policies for weather data partly because of pressure from the weather derivatives industry (Bates & Goodale 2017).

The broader argument is that the power of information lies in the ability of firms – and by extension, the nations in which they are located – to benefit from being able to manage weather-related risk. The irony is that this risk-attenuation may come at the cost of forestalling action about broader climate change. Insurance may insulate large firms and nations from negative weather impacts linked to adverse climate impacts. In extreme cases, powerful interests in information-rich nations may seize opportunities to capitalize on climate change's financial and weather turbulence. These opportunities conflict with the need to remedy global environmental change's underlying social and environmental problems (Cooper 2010).

5.4 Focus: Data Science for Sustainable Development?

Many scholars and policymakers increasingly position data science as integral to sustainable development, but challenges remain. A growing body of data science supports many facets of development, urban planning, and resource conservation. As Sarah Williams puts it, this works asks how we can use data as a tool for empowerment and not oppression (2020). At the same time, development is the site of many digital divides. *Sustainable development* is defined as that which balances the needs of the present and future generations, per the report *Our Common Future* by the United Nations World Commission on Environment and Development (1987). Development and sustainability are contested notions, as are many of the existing definitions of the regions where development can or should occur. There are exciting examples of data science supporting sustainable development and counter-examples of failed efforts. Data science must balance the promise of new data sets and advanced analysis with the dangers of instrumentalism and solutionism. These worldviews often ignore the lessons of what kinds of sustainable development appear to work well in reality.

Development is not a settled concept. There is a spirited conversation around what terms to use: developing, third world, Global South, low- and middle-income countries, among others. The term *developing* carries potentially paternalistic connotations of undeveloped (bad) and developed (good) nations. It also implies a given trajectory and denies how most countries have a range of experiences and goals. Anyone living in a developed country like the United States can see millions living in poverty and unsafe conditions. The United Nations Development Program allows countries to declare themselves developing to confer trade advantages. *Global South* has the benefit of not attaching potentially pejorative adjectives like "developing" to places, but the geographic moniker paints with a broad brush by capturing a mix of regions. The *third world* is out of date, referring as it does to a Cold War political situation. We partially sidestep this larger conversation by using *low- and middle-income countries* (LMIC). It relies on a quantitative measure, income, that maps onto many characteristics while recognizing that any income measure misses many vital features of these countries.

Sustainability is like development in being difficult to define. The core concept is value-laden because it assumes certain kinds of human-environment relationships, poses ethical quandaries over intergenerational equity, and can embody varying and sometimes competing moral stances (Mair et al. 2018). Sustainable is also complicated because what is sustainable in one sector may not be so in another, such as balancing economic growth with environmental protection. Tradeoffs are evident in one of the most commonly used frameworks for pursuing sustainability: the *Sustainable Development Goals* (SDGs). These are 17 goals broken into 169 specific targets tied to a suite of 230 underlying indicators (Figure 5.2).

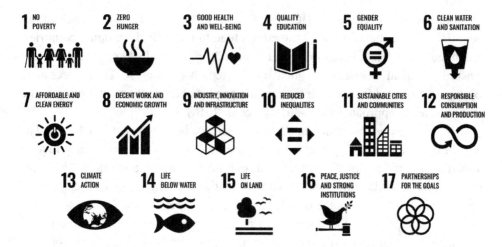

Figure 5.2 Sustainable Development Goals (United Nations 2020). Used with permission. The content of this publication has not been approved by the United Nations and does not reflect the views of the United Nations or its officials or member states.

The SDGs have a long history and resulted from the work of many people and organizations. They originate in part with the Millennium Development Goals, which had sixty indicators that monitored the twenty-one targets for eight goals defined by eradicating extreme poverty and hunger; achieving universal primary education; promoting gender equality and empowering women; reducing child mortality; improving maternal health; combatting disease; ensuring environmental sustainability; and developing global partnerships for development (Sachs 2012). The date for achieving these goals was 2015, and it is fair to say that while progress was made, especially in terms of disease and maternal and child well-being, the goals were not met (Lomazzi et al. 2014). The SDGs were meant to be more integrated and cohesive than the Millennium Development Goals because to achieve one goal it was necessary to achieve other ones, given how social and environmental systems are entwined (Thinyane et al. 2018).

5.4.1 Data Science for Development

Data science has answered the call for sustainable development in various ways. The field of *computational sustainability* centers on providing data and methods for policymakers and other stakeholders to make complex decisions about natural resources. These decisions should capture a range of human-environment dynamics (Gomes 2009). Other terms include information and communication technology for development (ICTD) (Dias & Brewer 2009), information and communication technologies for development (ICT4D) (Toyama 2010), or big data for development (BD4D) (Hilbert 2016). These efforts tend to focus on developing countries, as defined by the United Nations Development Program, and address problems that concern a critical development area for the region of interest. They necessitate solutions that differ from currently existing or plausible solutions in having a role for data science approaches (De-Arteaga et al. 2018).

There are reasons for optimism about data science helping to achieve SDGs. The amount of data digitally generated is on the rise, and these data do not require human intervention and its corollaries around time, training, and effort. They also promise continual data monitoring with the spatial and temporal specificity necessary to capture the 230 indicators that undergird progress toward the goals. Data-intensive projects have the potential to put actionable social and environmental information in the hands of the public in impoverished nations. The hope for systems like the World Bank Development Impact Evaluation initiative (DIME) is that artificial intelligence can be coupled to large data sets to measure and promote sustainable development and help target underserved populations (McDuie-Ra & Gulson 2020).

As noted in Chapter 2, there are ongoing discussions around the advantages and disadvantages of big data relative to more traditional kinds, primarily administrative data. Many LMICs use out-of-date or infrequently gathered socioeconomic measures (Devarajan 2013). These nations face challenges stemming from low administrative capacity, inadequate funding, and ill-coordinated outside organizations playing out against political sensitivity around data collection. There is also legitimate concern that official statistics are biased and incomplete, and the hope is that big data gathered from nongovernmental sources offer a different view (Sandefur & Glassman 2015). "Cash-strapped, infrastructure-limited national data systems run by staff who lack training and authority are common among poor countries. They are the biggest barrier to achieving the Sustainable Development Goals (SDGs) ... none of these goals can be met without a data revolution" (Espey 2019, p. 300). Of course, private firms and other data sources have their own biases regarding development statistics, just like other data from these sources (Kirkpatrick 2012).

Mobile technology, in particular, is a promising source of big development data. Cell phones are a lifeline for many people by offering banking services or access to up-to-date agricultural market and weather information. For example, mobile banking is credited for increasing financial system inclusion to 75 percent of Kenya's adult population in 2016, up from 26 percent a decade before (Ndung'u 2018, p. 44). However, it should be noted that there are still social and geographical divides in how the technology rolls out (Bernards 2022). Another example is how organizations like Question Box use technology to engage with illiterate populations, where people can speak questions on phones and get answers regarding health, agriculture, and commerce (Gibreel & Hong 2017).

Data science approaches, especially machine learning and other forms of artificial intelligence, promise more significant access to expert advice. Many regions struggle to find or keep laboratory technicians, agricultural experts, or disease specialists (Quinn et al. 2014). Education is another area where machine learning could significantly improve the use of less-known languages, one-on-one tutoring, and learning tailored to individual students (De-Arteaga et al. 2018). Medicine is among the most commonly raised examples because specialized medicine is unavailable or expensive in many parts of the world. Early diagnoses, tailored care, and expert interpretation of scans are the kinds of tasks that machine learning could address. The same rationales extend to public health when planning vaccination campaigns or dealing with infectious diseases (Tatem et al. 2009).

Having discrete SDG facilitates diving into discrete combinations of goals and data science approaches. It helps to lump these goals into the three loose categories of society (1–7, 11, 16), economy (8–10, 12, 17), and environment (13–15). Of course, these are all entwined; for example, it is hard to imagine making much progress on the socioeconomic goal of reducing inequality without attendant progress on education or gender equality. Similarly, their interdependence is laid bare by COVID-19 or other globe-spanning events. Some pathways to achieving SDGs are contingent on sustained economic growth and greater globalization, two trends that were slowed by the pandemic (Naidoo & Fisher 2020). A canvas of artificial intelligence and development experts contends that artificial intelligence would help accomplish 134 of the 169 targets but could also simultaneously inhibit progress toward 59 of these 134 (e.g., artificial intelligence promises to help alleviating poverty by helping target beneficial governmental economic interventions while also increasing poverty for some by eliminating jobs via automation). In addition, there was consensus on the need to focus more on the need for regulatory oversight of artificial intelligence in development to safeguard standards around transparency, safety, and ethics (Vinuesa et al. 2020). Section 5.4.2 delves into these and related issues.

Social SDGs center on advancing core facets of human well-being. They relate to: (1) reducing poverty; (2) ending hunger; (3) promoting health; (4) offering quality education; (5) ensuring gender equality; (6) providing clean water and sanitation; (11) creating sustainable cities and communities; and (16) developing peace, justice, and strong institutions, including those that support the SDGs (17).

Goal 1 is ending poverty in all its forms everywhere, focusing on eliminating extreme poverty in particular. Data science can play a role. For example, machine learning can use remotely sensed imagery to predict poverty as proxied by house-hold consumption and assets in lower-income countries (Jean et al. 2016). This work builds on a long history of scholarship examining how information technology can be used to help economic development, such as providing better pricing and weather information to farmers and fishers (Jensen 2007). There is emerging evidence that data science can support smallholder and subsistence farming, agriculture that provides a vast amount of food and livelihoods to the world's poorest regions. Successful cases include using data science approaches to offer better agricultural insurance, increase access to financial credit and payments, and link smallholder agriculturalists to export markets (Protopop & Shanoyan 2016).

The goal of ending hunger, achieving food security and improved nutrition, and promoting sustainable agriculture (Goal 2) is closely related to Goal 6, ensuring availability and sustainable management of water and sanitation. For example, cell phone data can inform models of cholera outbreaks in the aftermath of disasters, with the usual caveats around how these data may not be fully representative (Bengtsson et al. 2015). For this work to be helpful, it needs to offer improvements over existing methods. The most promising research centers on tasks that data science does well, such as predicting where famine or hazards may occur more quickly than current approaches (Kamilaris et al. 2017). Data science is expanding into many aspects of agriculture, including crop management and prediction of yield and quality; detection of diseases, weeds, and pests; livestock management related to animal welfare, production, and yield; and environmental stewardship, especially of water and soil (Liakos et al. 2018).

Goal 3 is ensuring healthy lives and promoting well-being for all ages. Health workers and policymakers have long identified the potential of new technologies to advance health care, including big data (Hansen et al. 2014). Artificial intelligence in automated health systems progressed in fits and starts due to psychological, technological, and financial impediments (Mishelevich 1986). The advent of machine learning and big data has reinvigorated the field of health informatics, especially the use of artificial intelligence to leverage health records (Combi & Pozzi 2019). There is also promise in low-cost approaches for disease detection via technologies such as cell phones or using machine learning to complement or even outperform human radiologists in reading X-rays and other imaging results for disease (Qin et al. 2019). Other tasks include identifying hot spots of vulnerability for people and the environment and linking the locations of services and populations at risk (Fleming et al. 2017) or surveillance systems for diseases, including COVID-19 (Hung et al. 2020). Data science approaches to health are promising in many contexts, including where public health infrastructure is lacking.

Ensuring inclusive and equitable quality education and promoting lifelong learning opportunities is Goal 4. Artificial intelligence approaches promise better learning experiences for a broader array of students. One such approach is the development of intelligent tutoring systems that are meant to supplement and perhaps even replace some kinds of instruction. However, further efforts must overcome challenges related to "student basic computing skills, hardware sharing, mobile-dominant computing, data costs, electrical reliability, internet infrastructure, language, and culture" (Nye 2015, p. 177). Although there is much promise in dealing with some of them, these are not small hurdles, such as when using machine learning to help develop content and real-time interaction in less commonly spoken languages (Sefara et al. 2019).

Goal 5 is to achieve gender equality and empower all women and girls. Data broken out by gender or sex in many places are unavailable, and only about a fifth of the data on the gender-specific SDGs is up to date (Rana 2020, p. 1). Big data – especially social media, mobility data, and remotely sensed imagery – can highlight issues of concern to women that are otherwise ignored or downplayed, including sexual violence, contraception access, workplace discrimination, and maternal health (Lopes & Bailur 2018). Some of these projects are data-oriented per se, as in collecting information related to gender equity. Others use data science tools to examine gender equity issues, such as when combing through social media or financial records to assess the state of women's equality.

Data science can help strengthen human settlements and institutions. Goal 11 is to make cities and human settlements inclusive, safe, resilient, and sustainable. Section 4.5 on smart cities explored the potential and pitfalls of using big data and data science to enhance urban living and environments, including those in the Global South. Promoting peace, justice, and effective, accountable, and inclusive institutions is Goal 16. This goal is related to Goal 17, strengthening the means of implementing and revitalizing the global partnership for sustainable development. One example of data science for building accountable institutions is using social media to track violence, crime, policing, and the legal systems (Chen & Neill 2015). Another example is using machine learning to examine procurement processes that are increasingly done electronically, and amenable to data science approaches, to identify and guard against inefficiency and corruption in public procurement (Gallego et al. 2021). Also promising is the use of machine learning to predict locations of election fraud using synthetic and actual data (Zhang et al. 2019).

The SDGs are centered on many traditional notions of development and therefore have a strong economic focus. These promote (7) affordable and clean energy; (8) decent work and economic growth; (9) advancing industry, innovation, and infrastructure; (10) reducing inequalities; and (12) responsible consumption and production.

Goal 7 is to ensure access to affordable, reliable, sustainable, and modern energy. A growing amount of research applies data science to the climate impacts of electricity generation and use. Electricity decarbonization and moving transportation away from carbon-based fuels are among the most critical ways of meeting the challenge of climate change (Williams et al. 2012). Current and planned uses include *load and source prediction*, guessing when and where demand will occur, and how and when a given source will be available. Data science is beneficial for assessing the capacity and timing of variable energy sources, such as wind, solar, or hydropower (Essl et al. 2017; Zhou et al. 2019) and power usage in complicated settings such as urban areas (Bibri & Krogstie 2017). Machine learning may also be helpful for seemingly mundane but essential tasks, like inspecting power lines and infrastructure to reduce transmission losses by applying machine learning to high-resolution aerial photographs (Han et al. 2019). While much of this work centers on energy systems, it can extend down to the level of individual buildings and rooms. Data collected by thermostats and motion sensors can be used to develop better models of energy use at the microscale (Al-Ali et al. 2017).

There are several entwined economic goals. Goal 8 is to promote sustained, inclusive, and sustainable economic growth, and full, productive, and decent work for all. Related are the goals to reduce inequality within and among countries (Goal 10) and ensure sustainable consumption and production (Goal 12). There has been much written on the potential for artificial intelligence to increase productivity (Acemoglu & Restrepo 2018). These increases threaten worsening inequality within and among nations if the benefits of this productivity flow to the few (Brynjolfsson & McAfee 2014). Big data can help promote transparency and accountability in policymaking in economic systems. Efforts such as the Billion Prices Project (BPP) use crowdsourcing and other forms of big data to provide alternatives to official statistics that are often massaged for political purposes (Cavallo & Rigobon 2016). There are flaws with these data, but even the flaws are interesting, such as showing where retailers inflate their online versus in-person prices because they know wealthier households – those with internet access – will pay more (Letouzé 2012).

Building resilient infrastructure, promoting inclusive and sustainable industrialization, and fostering innovation are the focus of Goal 9. Data science drives the growth of ride-sharing apps, bike and scooter sharing, and autonomous vehicles. Evidence is mixed on whether particular modes, such as ride-sharing, will create new problems like induced demand alongside new solutions (Milakis et al. 2017). The United Nation's Global Pulse initiative offers compelling examples of using artificial intelligence and big data to address infrastructure sustainability. Founded in 2009 at the direction of then Secretary-General Ban Ki-moon, this initiative centers on developing a global network of labs. These facilities bring together scholars and others to use free and open source computing approaches to address a range of development and humanitarian issues, focusing on sustainable infrastructure (Kirkpatrick 2012).

Sustainable futures necessarily involve humans interacting with the environment in sustainable ways. This form of sustainability includes (13) climate action; (14) protecting life underwater; and (15) safeguarding life on land.

Meeting the challenges of climate change may be the most pressing policy issue facing humanity (Goal 13). There is interest in using big data and machine learning for climate change detection, mitigation, and adaptation. Climate modeling exemplifies how researchers use computers to understand and predict a complex coupled human-environment system (Section 4.2.2). Efforts include using large data sets, tuning models, and helping choose among models when combining multiple models in ensembles (Anderson & Lucas 2018). Machine learning is also used to inductively approximate computationally intensive models from data, making model results more readily available to decision-makers (Scher 2018). At the interface of short-term weather and long-term climate, big data is increasingly used to predict extreme events such as tornadoes, cyclones, droughts, and floods (McGovern et al. 2017). Interesting work is being done on using cell phone records to track migration and mobility in climate-stressed regions to capture subtle shifts as people respond to changing habitability due to extreme weather events (Lu et al. 2016). Many other examples of using data science to measure and manage environmental systems appear throughout this volume.

Conserving life on oceans and land are Goals 14 and 15, respectively. These goals involve conserving and sustainably using the oceans and marine resources while protecting, restoring, and promoting sustainable use of terrestrial ecosystems (König et al. 2019). There has been a surge of interest in using data science approaches with remotely sensed imagery (Section 2.4) to address the SDGs (Holloway & Mengersen 2018). Remotely sensed imagery is a bedrock component of developing the global indicators framework necessary to implement the SDGs. The spatial coverage and temporal frequency of satellite observations outclass most other kinds of data collection and can complement other data sources (as explored in Sections 2.2.4 and 2.4). Much of this work is on land cover assessment, such as linking forest cover to ecological services, or increasing our knowledge of the oceans (Paganini et al. 2018). There is also compelling work on using machine learning to track human-environment interactions, from the growth of urban slums to water quality to deforestation fueled by illegal drug operations (Kugler et al. 2019). Data science is helping fill critical knowledge gaps about landscapes and the oceans.

5.4.2 Solutionism and Blind Spots

Data science for development often takes an instrumentalist focus that can veer into blind solutionism. Computational sustainability tends toward instrumentalism (Section 1.5) because it sees technology as value-neutral and under human control. In this view, "[n]o technology, including Big Data, is inherently good or bad for development" (Hilbert 2016, p. 156). As data science grapples with development, however, it has become increasingly clear that computation is only partially successful in advancing SDGs because it often fails to move beyond the general rhetoric of using data science to help people and the environment (Fadiya et al. 2014). Solutionism is the belief that most problems can be solved with just the right kind of data and algorithms while ignoring complex human-environment dynamics on the ground.

Data science for development suffers from shortcomings of *solutionism*. This worldview treats "complex social situations either as neatly defined problems with definite, computable solutions or as transparent and self-evident processes that can be easily optimized – if only the right algorithms are in place!" (Morozov 2013, p. 5) Solutionism is related to how data science and other fields need better integration. As explored in Section 4.2, data science can be incorrect or inapplicable to a given situation because the underlying data do not capture reality or existing knowledge well.

If we gather lots of (mostly well-educated male) programmers, armed with expensive machinery, and put them in a room with a tank of coffee, their version of "social change" will almost always involve finding the right open data set and hacking the crap out of it. Not only does the hackathon reify the dataset, but the whole form of such events – which emphasize efficiency and presume that the end result, regardless of the challenge at hand, will be an app or another software product – upholds the algorithmic ethos.

(Mattern 2013, p. 8)

In laying out the case for computational sustainability, Gomes (2009) focuses on computational challenges, such as determining the most optimal arrangement of land corridors to preserve habitat for migratory species or managing catches in fisheries. This work is important and can contribute to solutions. However, the problem is usually not one trying to weed out slightly less optimal solutions but instead developing feasible political and economic regimes. "Too often, researchers publish exciting results about the application of AI to an environmental problem and those results are never translated into applications. Or an AI system is handed to a non-profit organization or government agency that lacks the resources and expertise to take advantage of it" (Joppa 2017, p. 326). In simple terms, many data science efforts fail by ignoring social and cultural norms, failing to build local relationships, or providing technology that is not appropriate to the sociotechnical context or capacity

Consulting companies and firms promulgate rosy data science futures for development that may not tell the whole story. These groups have a vested interest in the public financing of private firms to serve technocratic interests. This focus is not necessarily a problem since it can be a successful development model. However, it is telling that many overviews of BD4D rely heavily on news releases and consulting reports and not scientific literature, with the caveat that this literature can take years to be published (e.g., Kshetri 2014; Letouzé 2012). People and places in LMICs can be seen as markets for experimentation with new products and capture, including hooking people on web-based services. These incursions may help propagate socioeconomic ills by providing antidemocratic governments more tools to surveil and control their populations or circumvent attempts to increase the capacity of civil society to provide new services (Arora 2016).

There is an uneasy alliance between data science and the larger development community. "ICT4D sprouted from two intersecting trends: the emergence of an international-development community eager for novel solutions to nearly intractable socioeconomic challenges; and the expansion of a brashly successful technology industry into emerging markets and philanthropy" (Toyama 2010, p. 12). By investing in newer systems that promise development via algorithms and development, there is a danger that aid institutions are pivoting away from the successful strategy of directly investing in the people and places in the Global South in ways that empower local actors: "the battle being fought in the AI and development space seems to be between the information expert and the development expert, the latter suddenly anxious that their future may be at stake or at the very least their in-country expertise devalued in favor of a new form of precision via algorithms and data-analytics" (McDuie-Ra & Gulson 2020, p. 629). While the US Agency for International Development and other large and traditional organizations have their critics, they are embracing data-driven projects in ways that seem better tied to work on the ground (King & Martin 2015).

The relationship between data science and sustainable agriculture exemplifies many of these tensions. Bringing more data and analysis to agriculture is good because more people can be fed, but it also perpetuates existing agricultural systems and creates new risks. Worldwide, major players, including Monsanto, Land O'Lakes, and DuPont, are moving into big data and collaborating with companies such as John Deere, a farm equipment manufacturer, to share data and expertise (Kshetri 2014). Agriculturalists in LMICs increasingly rely on data-oriented tools, such as cell phones and text messages, to convey crop, input, weather, and insurance information. These efforts face hurdles in how adaptive capacity remains low and tied to microscale social and environmental challenges that data science can only partially address (Lwasa 2015). Overall, these companies offer economies of scale and efficiency that promise to benefit farmers and consumers.

Data science for agricultural development gives rise to concerns. There are fears of a "data land grab" because agriculture is increasingly data-intensive. There is a real threat of dispossession of data (Section 5.1.2), where farmers provide their data to an upstream producer but do not ever really own these data (Fraser 2019). For example, John Deere and other equipment manufacturers limit how consumers or nondealers can fix equipment or access data collected by this equipment. Farmers worry that these companies are well-positioned to manipulate markets or products using their data sets on climate, yields, and input use (Wiseman et al. 2019). Data science can increase dependency on corporate technological systems, unsustainable expectations about productivity, and greater market and service consolidation (McCarthy 2016). For example, an Indian industrial conglomerate has moved into agricultural technology and web applications for farmers to make data-driven decisions. The company, which "also owns India's largest supermarket chain and India's largest telecom services, could conceivably gain complete control of supply and demand chains – with farmers relying on the app to sell their products via the company's telecom service directly to the company's retail stores" (Parida & Ashok 2021, p. 163).

Like any other development strategy, data science solutions do not work when they ignore broader social and environmental contexts. Solutionism can create profound disconnects between optimistic but short-term technological fixes and the more profound and longer-term engagement necessary to effect change (Maalsen & Perng 2016). High-profile projects like Data for Development (D4D) bring together a range of experts to work on big data to help spur development in low-income countries, in this case, Côte d'Ivoire. The resulting papers on population, economics, and health were impressive. However, it produced little engagement with the people on the ground or the policymakers in a position to effect change. The project primarily enlisted data scientists alongside mathematicians, physicists, and others who work with big data but far fewer policy experts or domain scholars in the social or environmental sciences. Similarly, the project did not engage with governmental bodies, and only one of the more than 200 teams visited the country (Taylor & Schroeder 2015). Successful projects cannot ignore social and cultural norms, fail to build local relationships, or provide inappropriate technology for a given sociotechnical context.

5.4.3 So, What Works?

So, what works best when applying the data science of human-environment systems to sustainable development? Technology for development generally works when it magnifies existing capacity or interests and fails when it is seen as a replacement for, or a simple addition to, this capacity (Toyama 2010). Data science is promising because it can quickly and inexpensively scale up. Complicated tasks for sustainable development, such as assessing forest cover with regular remote sensing or modeling riverine flood potential, can be inexpensive once the basic technology is in place. Increasing nontechnological capacity is usually more expensive. Doubling the number of health care workers or agricultural extension agents can require more than doubling the expenditure on people and supportive infrastructure. More broadly, technology as a multiplier works better for people on the wealthier side of the digital divide, those with languages, networks, money, time, privilege, or other attributes that ease their access to and use of technology. Applying it to other contexts requires nuance.

A truism in development is that people can know what to do but not have the capacity to do it. Data science for development can miss this point when assuming a critical roadblock to development is a lack of information or analysis. Of course, sometimes this information is missing and is incredibly helpful when developed. There are data divides in research in that "the gap between academia and public policy, which is often wide, is especially so in lower-income countries where there is often no access to academic publications, even if policymakers were aware of the project and wanted to read its results" (Taylor & Schroeder 2015, p. 509).

Other times, new or additional information may not have the desired effect. For example, using locational information from cell phones can help model the growth of slums, which "could help the government to optimize resource allocation for infrastructural development and other resources" (Kshetri 2014). While there is genuinely exciting potential with these sorts of approaches, it is not always clear how these data add anything that could not be accomplished by talking to locals who have a good idea of where future growth will occur. Moreover, this knowledge may be moot if there are no resources to accommodate growth. Joining multiple forms of data and analysis seems to offer the best path forward for research, but it must be aware of the broader context.

Data science approaches promise to create data on underexamined topics and places. There are deep data disparities between high- and low-income parts of the globe. For example, there is little systematic data on low-income countries' social and natural features. They may lack primary forms of data collection that have long been common in high-income countries, including censuses, meteorological measurements, and land surveying. As a result, there are gaps in the data underlying SDG indicators, including those related to child and maternal mortality, household income, and many features of women, minorities, and indigenous populations. Big data have great potential to fill these gaps, but they are often tied up in private hands, such as cell phone activity or social media (Swain 2018). These data can be very hit or miss when offering well-crafted social and environment variables compared to other forms of data and pose a range of privacy challenges (Letouzé & Jütting 2015). Nonetheless, they may end up being the best data available.

Private–public partnerships and open data hold some promise as a way forward in bridging data divides. For example, as noted earlier, data science for agriculture offers challenges and opportunities. There is work in developing agricultural data codes of practice that could apply to many regions, including the American Farm Bureau's Privacy and Security Principles for Farm Data and New Zealand's Farm Data Code of Practice. These efforts seek to establish and regularize issues around data, including consent, disclosure, and transparency (Sanderson et al. 2018). There are also advocates for more open data and modeling ecosystems (Antle et al. 2017). Open systems could address weaknesses in existing agricultural models, which are numerous, redundant, poorly integrated, and rarely reused. The profusion of data sources gleaned from on-ground observations and ease of accessing models in the cloud could spur a revolution in open agricultural modeling.

Also essential to successful development are collaborations among domain experts, data scientists, and stakeholders. Efforts like the Food Security Portal, facilitated by the International Food Policy Research Institute (IFPRI), provide data-driven expertise on food security and its interaction with global food systems (Trethewie 2021). It relies on research to balance data quality with timeliness and relevance. The project relies on domain experts who draw on a range of data, including real-time reports and futures-trading reports, to provide firm estimates of current and future events. It remains to be seen what potential exists for yoking big data to enhance this work, but there are the makings of a strong beginning. What if data science can give a week's notice, or a month's notice, of a new crisis? How many lives could be saved, or how much human well-being safeguarded? Actual impacts may be slight if the underlying real-world systems cannot use these data, but they may be a solution if and when they can. For example, the portal offers a food security mapping service driven by Google news updates. This system takes advantage of elements of big data, including using real-time data sources drawn from a vast trawling effort. It is not hard to imagine extending this work to related progress on automating sentiment analysis, interpreting remotely sensed imagery, and drawing on crowdsourcing (van Ginkel & Biradar 2021).

Data science and technology can be purposely created for sustainable development. Many data science approaches are created with high-income or high-capacity countries in mind in ways that can limit their utility for other contexts. However, computer science, statistics, and other fields are also more than capable of designing hardware, software, and methods for low-resource settings (De-Arteaga et al. 2018). For example, there is much effort around developing low-powered cell phones, satellite-based internet, and many analytical routines designed to deal with noisy or conflicting data sets (Section 3.3.5). Data science is focused on technical issues over social ones:

Can mobile phones provide income generation and facilitate remote medical diagnosis? How can user interfaces be designed so they are accessible to the semiliterate and even the illiterate? What role can computers play in sustainable education for the rural poor? What new devices can we build to encourage literacy among visually impaired children living in poverty? What will a computer that is relevant and accessible to people in developing regions look like? *(Dias & Brewer 2009, p. 74)*

There is a focus on software and hardware being designed for people in LMICs, with an overriding sense that it is part of the larger wave toward ubiquitous computing. There is also some focus on sustainability couched in financial terms, where commercial success may help new systems take root and flourish. Again, this work tends toward solutionism, but it may lead to good solutions.

Beyond technological advances, successful data science for development depends on policy context. For example, in Brazil, India, and China, combinations of three different policy visions are at work around data science and development (Mahrenbach et al. 2018). The first vision sees data as either a force for liberation or repression depending on how invested the government is in controlling citizens versus seeing data as a way to improve democratic participation and freedom of expression. The second sees data as a means of improving government service provision by identifying gaps and opportunities regarding the efficiency and effi-cacy of these services. The third situates data to promote social and environmental development, including using technology to improve socioeconomic well-being and as a vehicle to offer economic incentives for environmental protection. In aggregate, all three countries have detailed plans concerning big data and have tasked specific public and private actors to carry out these plans, although the effectiveness of these plans remains to be seen.

There is also growing recognition of the need to work with domain scholars and people outside of technology. "Sociologists, ethnographers, and anthropologists, for example, can provide valuable information about the communities intended to benefit from ICTD. This information, regarding such things as cultural practices, traditions, languages, beliefs, and livelihoods, must guide the design and imple-mentation processes for successful solutions in ICTD" (Dias & Brewer 2009, p. 76). The authors also note how political scientists and economists should be involved in designing new marketing strategies and economic models, along with a range of practitioners and policymakers. There is also a real need for social scientists and other scholars to weigh in on the use of big data for developing indicators for SDGs. Many seemingly simple indicators can be seen as reductionist and gloss overcomplicated and contested concepts because they impose a veneer of objectivity on subjective concepts like equity or well-being. The value of these indicators is helping make real the goals of SDGs and providing a way to measure progress (or lack of it), but to fully implement and understand indicators requires that critique is heeded and integrated (Mair et al. 2018).

Crowdsourcing disaster relief highlights many challenges and opportunities in linking ground knowledge, domain knowledge, and data science. There is much good in using data science and crowdsourcing as part of humanitarian responses to disasters, but such aid only goes so far. *Digital humanitarianism* is the use of big data trappings, such as spatiotemporally rich satellite imagery, social media, camera-equipped drones, and machine learning to aid humanitarian responses during disasters such as the 2010 Haiti earthquake (Meier 2015). Communications-centered projects like Mission 4636 allowed Haitians to report via text messaging incidents and needs. The Ushahidi Haiti Project had volunteers from around the world map information from messaging, email, and web-based forms. People also contributed to OpenStreetMap, an open source map to which anyone can add information. Volunteers from around the globe added data for roads, buildings, and other features to create one of the most detailed maps available, and many people on the ground adopted it.

As in other settings, there were noticeable gaps in access to the technologies and who benefited. For example, messages were translated to English, limiting their use by many native Kreyol speakers, and many people in the country lacked the necessary internet access to contribute to or use the data (Crawford & Finn 2015). Digital humanitarianism can help develop situational awareness but may not be particularly helpful in guiding aid in the short term. At the same time, longer-term efforts suffer from data science's potential to reify preexisting social or political relationships and inequalities (Burns 2015). While crowdsourcing offers new resources, emergency response officials did not necessarily trust the information because they knew that not everyone had access to cell phones, which was necessary to report information. Overall, Ushahidi was valuable to a range of stakeholders, offering relevance, efficiency, effectiveness, and impact, especially in how it was often the only source of information for situational awareness in the early aftermath of the earthquake via the spatial aggregation of data (Morrow et al. 2011), with caveats.

In sum, development challenges are rarely just about needing more knowledge or analysis of the kinds that data science provides. Roadblocks to development are more often tied to insufficient social, economic, and political resources. As a result, data on their own are not the panacea that they are often portrayed to be. A truism in social science research is that most natural disasters are not natural but expressions of socioeconomic and political vulnerability (Blaikie et al. 1994). The same forces biasing data also guide predisaster and postdisaster socioeconomic structures and dynamics. More rural, older, and less wealthy areas are less likely to have access to technologies underpinning big data (e.g., smartphones, broadband internet) and these gaps often mirror broader societal ones in the ability to prepare for or respond to extreme events.

Data science for development is promising but must be applied with knowledge of the larger context. It is critical to involve people and organizations on the ground at all phases of research. This work includes participating in defining and collecting the necessary data and not assuming it will come from big data in the cloud (Thinyane et al. 2018). Community-based organizations play a critical role in offering services to the local population while also working with these communities to facilitate participation by traditionally marginalized groups. These groups serve as a conduit and as a way to prevent undue exposure of private data or other harms that can come from data collection, dissemination, and policies stemming from these data.

5.5 Conclusion

Data science applied to human-environment systems have many policy dimensions. Data science can advance policy in some realms, like helping protect natural resources or root out discrimination. At the same time, it can contribute to problematic issues around privacy, dispossession, and surveillance. Policies designed to make data science more effective or less harmful to people and the environment form a crazy quilt of jurisdictions and lag behind many of data science's technological and societal aspects. Positive developments are occurring, and human-environment scholars are gaining more tools to build policy-relevant considerations into their work. However, much remains to do in building the scientific, legal, and policy instruments that facilitate open science and address social and environmental impacts. Ongoing work in sustainable development is helping advance the conversation around extending data science and human-environment research to a broad range of human and natural systems in ways that help people and the environment where most needed.

6

Ways Forward for the Data Science of Human-Environment Systems

Data science plays an increasingly prominent role in examining and addressing human-environment dynamics. The growing role of data-intensive inquiry is apparent to many researchers, policymakers, and others. At the same time, however, there is mounting awareness of the interplay between the perils and promise of big data for human-environment systems in particular. Despite all the research and interest, there are few straightforward ways to reconcile the pros and cons of data science. This difficulty stems partly from the many unknowns about data science because its story is still being written. There are exciting developments in data science theory, conversations around democracy and decision-making, and promising developments in data science infrastructure.

6.1 Data and Theory for Science

Poovey (1998) offers a fascinating account of how numbers were used in science and commerce and gave rise to the *modern fact*. This notion, that numbers and facts speak for themselves or are independent of a larger social milieu, is relatively recent. Numbers are not just numbers, and by extension, data are not just data. They are signifiers of a more extensive system objectively layered with knowledge and trust. In the era of data science, scholars are rediscovering the fundamental principle that many numbers are not standalone entities but are embedded in complicated political and socioeconomic dynamics. Many of the debates about the conceptual basis of data science are facets of broader epistemological fissures in science.

As big data are increasingly embedded into social and spatial decisions, processes, and institutions, the links between signifier and signified might become ever more obfuscated. That is, as we begin to place more and more trust in big data and the software, algorithms, and machines that are used to produce and analyze such data, we may tend to lose sight of the very things that such data represent, blurring the boundaries between the ontological and epistemological. *(Graham & Shelton 2013, p. 259)*

There is room for more fruitful conversations among data science, human-environment research, and science and technology studies around the constructionist precepts of data. We need more conversation between people who see themselves as "doing the work" of data and those who engage with the more abstract questions around data. One way forward is learning from ethnographies of data projects, which seek to explore and lay bare the messier yet central social, economic, and cultural facts of data capture. For example, Walford (2018) explores a data collection project in the Amazonian forest, laying out the many pieces involved in collecting data and high-lighting a profoundly complicated set of connections around awareness, power, exploitation, goodwill, and ignorance that define the various people necessary to collect data. These dives into scientific practice make concrete some of the more abstract claims around data science.

Another way forward is for scholars of all stripes to reengage with work on the relationships among data, method, and theory in research. While many data science practitioners ignore existing domain work, many social and environmental scientists ignore foundational work in data science, seeking to redo or reimplement systems and approaches or ignoring statistical best practices (Section 4.2). For example, much duplicative and standalone work exists in the social and information sciences on crowdsourced data (Graham 2018). This engagement allows researchers to collectively and collaboratively assess the challenges and advantages of big data while also clarifying the role and value of existing scholarship (O'Sullivan & Manson 2015). Given how scholarship seems to be narrowing and becoming more specialized over time, there is a growing role for infrastructure and large projects to help spur these broader conversations, as Section 6.3 explores.

Human-environment research helps make data science better because it is interdisciplinary and grounded in real-world problems. In addition to data scientists jumping in, there has been a proliferation of social scientists interested in the environment, and environmental and physical scientists interested in human systems. This growth reflects recognition that interdisciplinary teams are needed to tackle complex research. The last three decades have seen the emergence and profusion of explicitly human-environment research programs with data and modeling as foci, including Future Earth, the Intergovernmental Platform on Biodiversity and Ecosystem Services, the International Geosphere-Biosphere Program, the Millennium Ecosystem Assessment, the Intergovernmental Panel on Climate Change, the Global Land Project, the Earth System Science Partnership, and Resilience Alliance. These efforts have data and analysis at their heart, and efforts like this can serve as an arena where data science plays out.

We increase our understanding of the human-environment system through inquiry, and data science will help in some ways and not others. As O'Sullivan puts it, "understanding the world still demands that we carefully develop theories, consider the implications of those theories for what we expect to observe in the world, and subject those expectations to scrutiny through empirical observation, using multiple methods, only a few of which are enhanced by the dragnet of big data collection" (2018, p. 24). It is impossible to have purely inductive or theory-free research because all interpretation is built on assumptions (implicit or explicit) that are built into how data are collected and processed.

Moreover, many scholars hold that we are not doing science if we halt the search for hypotheses and models, especially by stopping at patterns and not looking for explanations (Pigliucci 2009). As Leonelli (2019) argues, many sciences have been practicing the core precepts of big data and data science for decades. Fields like astronomy and meteorology have long collected, managed, and visualized large and complex databases. The growth of data science is the next phase of a centuries-long evolution in how scholars employ data to increase understanding of the human-environment systems. We have to trust that scholarly truth will win out while accepting that there always be tensions and complementarities among different modes of science around truth-making.

6.2 Data Science, Democracy, and Decision-Making

Research may or may not inform wise policymaking. While human-environment scholars have different rationales for doing their work, many are interested in making the world better and see their work as a way to get there. The corollary of this view is that many researchers see poor or incomplete knowledge of the world as the primary roadblock to better decision-making. The data and decision-making relationship is expressed in many ways, but one of the most common is the data-information-knowledge-wisdom (or DIKW) pyramid. Of course, the link between decision-making and scientific research is not especially strong. Approaches like data science join other forms of inquiry in grappling with how research is embedded in a social and environmental context. In particular, as humans increasingly weave data and algorithms into their decision-making, these digital representations become inseparable from human and environmental systems.

The DIKW pyramid is a view of how research guides policy. It is attributed to a number of different scholars and artists for half a decade and has been codified in the years since (van Meter 2020). The pyramid is used implicitly and explicitly in data science and human-environment research to think about translating research into action (Figure 6.1). This schema rests on data about various aspects of a subject or topic. The information tier captures critical elements of a subject, while knowledge is

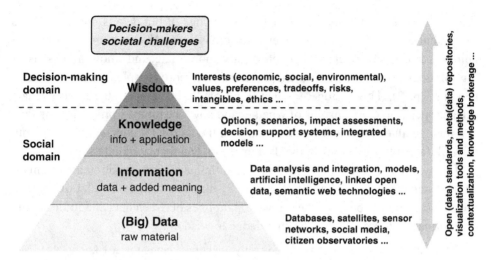

Figure 6.1 Data–information–knowledge–wisdom pyramid for human-environment challenges (adapted from Lokers et al. 2016). Reprinted with permission from Elsevier.

concerned more with what we know in an integrated way. Wisdom is the capacity to act reasonably based on data, information, and knowledge (Rowley 2007).

Moving from data to information is complicated by the existence of different kinds of information. Buckland (1991) distinguishes three kinds. Information-as-process is the act of informing a person or workflow, increasing their stock of knowledge. Information-as-knowledge is a new fact, subject, or event conveyed via information-as-process. Finally, information-as-thing maps onto the artifact of data, the document, text, or measure. It may also correspond to a physical or digital artifact. This latter view distinguishes between knowledge, which is essentially intangible and unknowable beyond the person who accesses it, and the information as an artifact that can be shared.

Taken to the extreme, this conception of data considers objects being measured, such as a stream or rock formation, as a kind of information. This notion of thing-as-information is easier to contemplate in the sense of a museum where birds, insects, or rocks are stored and cataloged. A rock is a form of information because it is held in a specific place and is the subject of metadata. This notion of objects-as-information is probably too far afield for most people concerned with data science. However, it helps illustrate that there can be fuzzy distinctions among objects, data, information, and knowledge.

Data science may help with the jump from knowledge to wisdom. With the advent of big data, Prensky (2009) argues that we are becoming *Homo sapiens digital*, or the digital human, and that the challenge of big data is not data versus theory, but one of wisdom. In this view, wisdom is defined by problem-solving: "the ability to find practical, creative, contextually appropriate, and emotionally satisfying solutions to complicated human problems . . . a more complex kind of problem-solving" (p. 2). This vision of a new kind of digital human supported by data science speaks to core tensions of whether and how one jumps from collecting data and making algorithms to effecting change in the world. Of course, there is plenty of work left in fixing issues in the data, method, and theory of data science.

It may be a while before these digital humans can tackle many human-environment problems. The DIKW hierarchy may not even make a lot of sense in the data science age because there few straight lines between data and information, or information to knowledge, or knowledge to wisdom.

Information, if accurate, is a more correctly identified input toward knowledge. Yet, inaccurate, misunderstood, misused, or false information is a real problem, and detracts from rather than increases knowledge. Knowledge is acquired information, skills, and education in a particular field or area, but is limited to that field or area. Knowledge requires information but also requires skills or education. Wisdom is intelligence applied with accumulated knowledge for the benefit of humanity and human environs, and further requires knowledge along with application of intelligence.   *(van Meter 2020, p. 77)*

The link between decision-making and scientific research is not as strong as many researchers believe. There is increasing recognition that scientific knowledge alone is rarely enough to change the world. It is necessary to incorporate human values, environmental ethics, and other sociocultural, economic, and political concerns into the use and interpretation of knowledge. We need data to make good decisions, but there is growing overreach and misuse, along with the recognition that effective governance has much more to do with socioeconomic power and politics than solid research (Alonso & Starr 1987). For example, to deal with error and bias (Section 2.3), there are proposals to move scholarship beyond simple notions of bias and be more invested in concepts of power. This perspective would account for historical socioeconomic inequities, labor conditions, and a fuller range of epistemological views inherent to data. This approach means moving "from bias research towards an investigation of power differentials that shape data" (Miceli et al. 2022, p. 3).

This view refines our understanding of how data have an inherent context of why they were collected. For example, many of our social and environmental data sets are increasingly commercial, and this orientation can build inherent biases in these data and any subsequent analyses. Bias moves from being a narrow technical issue that can be solved by better sampling or algorithms toward recognizing that "unbiased" data can still fully reflect the power imbalances in the social phenomena being described. It requires much work to unearth how these data relate to the real world (Gebru et al. 2021).

Individuals, groups, and policymakers regularly make decisions absent science or despite science. Climate scientists made headlines in 2021 when some called for a strike:

> The science-society contract is broken. The climate is changing. Science demonstrates why this is occurring, that it is getting worse, the implications for human well-being and social-ecological systems, and substantiates action. Governments agree that the science is settled. The tragedy of climate change science is that at the same time as compelling evidence is gathered, fresh warnings issued, and novel methodologies developed, indicators of adverse global change rise year upon year ... Given the urgency and criticality of climate change, we argue the time has come for scientists to agree to a moratorium on climate change research as a means to first expose, then renegotiate, the broken science-society contract.
>
> *(Glavovic et al. 2021, p. 1)*

While, on the surface, this call for the cessation of some kinds of research seems extreme, the authors are correct in how doing more science, even with new data science tools, is unlikely to occasion some new flowering of coherent policymaking.

Another challenge is that technology is not value-free and is firmly embedded in a social and environmental context that plays a significant role in how data science works. Data science promulgates a blend of instrumentalist and constructionist visions (harkening back to the philosophy of science discussed in Section 1.5). However, constructionism seems to explain how data science plays out in many places. Technology is shaped by various human-environment forces that mold many aspects of nature and society. As such, the impact of data science is contingent on the broader social milieux in which it plays out. In some cases, solutionist and instrumentalist data science can harm people and the environment, and in others, it is just ineffective. At the same time, data science and big data may offer new solutions to both new and old human-environment challenges and give rise to optimism. Data science may offer openness and democratization in some regions and better data-driven policy in others, as explored for sustainable development in Section 5.4.

The potential for success in moving up the DIKW pyramid may lie in shifting how data science is situated in a broader social milieu. For example, crowdsourcing is ushering in a new age of democratizing data, decision-making, and policy. The simplest form of democratization is the idea of more people having access to a specific form of technology, such as global navigation systems, enabling locational awareness for projects like eBird, iNaturalist, or Zooniverse that have "helped to foster an unprecedented democratization of geographic knowledge, often with roots far removed from academic experts" (Warf & Sui 2010, p. 200). Citizens can use spatial data and mapping to influence public discourse and decision-making around local to regional issues, participate in weather and climate observations, or link backyard animal observation to climate variability and change (Sheehy 2019). They can make visible what was previously unseen and contest accepted narratives promulgated by those in, or with, power, such as entrenched political or financial interests (Muñoz et al. 2020).

Then again, much crowdsourcing tends to be instrumentalist in orientation, focused more on the data being collected than the subsequent processes by which data are not just collected but used. As such, this work can suffer from the delusion of democracy, a superficial exercise that is valuable in many ways but ultimately does not have a considerable impact (Haklay 2013). Responding to the broken science–society contract, and by extension, issues around democratizing data science, requires work that engages more with politics of environmental knowledge and is better at understanding obstacles to action (Turnhout & Lahsen 2022).

Finally, there is an entirely warranted and ironic concern about how a data science of human-environment systems can mold these systems in poorly understood ways. By relying on data science and big data, humans create complex and potentially unknowable models that guide and predict decision-making (Weinberger 2011). At issue is that they can go beyond prediction and shape our reality in ways that lie beyond human comprehension and perhaps even control. In terms of the DIKW pyramid, the bottom data layer may come to dominate, as decision-making is short-circuited and humans are left behind.

In this situation, causality becomes something beyond just a modeling exercise. "Big Data may appear to successfully understand and predict things, but only because powerful interests behind them are shaping the world … if Big Data pick up trends, and are then used to facilitate those trends, then the data cease to be a tool for understanding the world but rather one for shaping it" (Shearmur 2015, p. 966). In essence, artificial intelligence is increasingly invisible to users and increasingly tasked to guide what people know and their choices in ways that promise to erode self-determination. The state of ethics that guides both developments in artificial intelligence and its use is woefully underdeveloped. There is much interest in charting ways toward ethical artificial intelligence, but much work is to be done (Taddeo & Floridi 2018).

6.3 Infrastructure and Interdisciplinary Projects

One of the most promising ways forward with data science is shared research infrastructure and large interdisciplinary projects. While interdisciplinarity can be time-consuming and frustrating, it supports mutual understanding around data, methods, and theory. It is also one of the best ways to advance knowledge of complex human-environment systems since no single scholarly domain can encompass these systems. This research is increasingly supported by centralized infrastructures that facilitate researchers working with data and performing analyses at a high level without spending much time or effort collecting, storing, and manipulating big data sets.

Infrastructure has key characteristics. Star and Ruhleder (1996) developed a set of dimensions via discussions centered on an early data infrastructure, the Worm Community System, that supported a network of over a thousand biologists studying nematodes. While created for a specific infrastructure, these dimensions are broadly applicable to other infrastructures (Table 6.1). Importantly, these characteristics emphasize that infrastructure is usually less about objects or people and more about the relations among them and how these relationships are mutually constitutive of these systems and people. Infrastructures "are not systems, in the sense of fully coherent, deliberately engineered, end-to-end processes" (Edwards et al. 2013, p. 6) but instead agglomerations of evolving systems comprising people, technology, and institutions. Technology is fundamentally social, and society has always been mediated by technology, especially if seen broadly as any process or artifact used by humans.

Researchers must unearth their infrastructure in ways that reconcile the larger goal of big data with the actual goals of researchers on the ground. For example,

Table 6.1 *Infrastructure dimensions*

Dimension	Description
Embeddedness	Ways that infrastructure ties into other structures, social arrangements, and technologies
Transparency	Infrastructure is transparent or even invisible once developed and supports its defined tasks
Becomes visible on breakdown	Conversely, infrastructure becomes visible or apparent to users when it fails to work somehow
Reach or scope	Most infrastructures exist beyond a particular spatial location or a single event in time
Membership learning	Knowing about the infrastructure requires learning and sustained engagement by members of a group that can be professional, social, or cultural. Outsiders see the infrastructure as something new, an object of inquiry that becomes learned as they become members of a group
Conventions	Linked with existing conventions and practices in often unrealized ways, as when inheriting a specific technological choice in new contexts
Embodying standards	Infrastructure embodies or adopts existing standards
Built on an installed base	Just as infrastructure embodies existing standards, it will be sculpted by existing infrastructure and guide future development

(After Star & Ruhleder 1996).

environmental sensing infrastructure is often envisioned as advanced machinery or vast sensor networks. However, it often makes more sense for researchers to have mobile and flexible sensors wielded by humans and supported by networked infrastructure (Mayernik et al. 2013). In simple terms, field-based environmental research is time-consuming. It requires highly trained scholars because a good deal of experience and knowledge goes into recognizing and adapting to the situation on the ground. Much of the data collection in the environmental sciences focuses on characterizing specific phenomena in small areas – surveying and collecting plants, counting animals or their signs (e.g., bird calls, sightings), or measuring physical features such as soil horizons (Section 4.3). This work extends into the lab using old and new methods, from weighing dried samples and chemical analyses to conducting DNA tests and taking microbial cultures. Infrastructure can link this fieldwork across people and sites.

The National Ecological Observatory Network (NEON) exemplifies how infrastructure can bring researchers and research together. The US National Science Foundation (NSF) funds NEON to network forty-seven terrestrial and thirty-four aquatic field sites. Researchers gather decadal-scale data on how the biosphere responds to land change and climate shifts, including interactions with the geosphere, hydrosphere, and atmosphere (Hampton et al. 2013). The network is intended to support spatial scales from individual sites to the entire nation at timescales from days to decades. This work is exciting, but there are ongoing conversations around what its data infrastructure looks like, especially in terms of logging and using data, strengthening data networking within and among sites, standardizing data collection, and sharing data with outside investigators and citizen scientists (Hinckley et al. 2016).

Other international networks join NEON in linking data across sites. The Critical Zone Observatory (CZO) network focuses on the *critical zone* of the Earth's terrestrial surface and has dozens of sites around the Earth (Guo & Lin 2016). The Terrestrial Ecosystem Research Network is an Australian program established in 2009 to capture and coordinate monitoring data at the continental scale for hundreds of field sites related to soil, water, carbon, vegetation, and biodiversity (Thurgate et al. 2017).

Projects like NEON and CZO integrate with others. For example, the AmeriFlux network was founded two decades ago and now encompasses 260 sites in the Americas that continuously observe carbon, water, and energy fluxes (Novick et al. 2018). The spatial, temporal, and attribute resolution of these data make them helpful in understanding relatively slow and long-term changes in climate and the land surface, and offering insight into events like droughts or flooding. The AmeriFlux observation system has been integrated into the CZO network, Long Term Ecological Research Network (LTER), and NEON.

Central to data science is the idea of cyberinfrastructure, which has been the focus of several decades of investment and growth as public agencies in many countries identified the importance of large-scale technical infrastructure centered on computing and the Internet (Parashar 2019). Another concept central to human-environment infrastructure is the *digital earth*, which centers on the idea of a computational infrastructure that measures and models the human and environmental features of the Earth. It also makes them available to scientists, policy-makers, and members of various publics (Craglia et al. 2012). Digital earth offers functionality to many users without specialized software, offering the most current and correct data, facilitating high-performance computing, offering access to existing model components, and sharing workflows (Hofer 2015). Large infrastructure projects in the Earth and environmental sciences embody these concepts at the forefront of a revolution in how many researchers do their work, as seen with efforts such as Earth Systems Grid, Environmental Virtual Observatory Pilot, or EarthCube (Cutcher-Gershenfeld et al. 2016). Centralized and online systems provide researchers access to data and tools.

Future Earth exemplifies infrastructure dimensions of conventions and embodying standards. This effort was proposed at the United Nations Conference on Sustainable Development (Rio+20). It was codified in the years following as an agenda-setting project driven by the participation of a dozen groups (Future Earth 2014). These include global-change programs such as the World Climate Research Programme (WCRP), the International Geosphere-Biosphere Programme (IGBP), and the International Human Dimensions Programme on Global Environmental Change (IHDP) (Uhrqvist & Linnér 2015).

Future Earth offers a broad array of research projects in addition to the Partnership for Resilience and Preparedness (PREP), which seeks to address the challenges of uncurated data, poor communication among data users and providers, and the need for data visualization and analysis (Future Earth 2019). There are related calls for a global Earth observatory with a thousand or more sensing stations scattered over the globe that would scale up existing regional networks. Such a network is the Pan Eurasian Experiment (PEEX), designed to fill gaps in our knowledge about the climate in the arctic and boreal pan-Eurasian regions and China (Lappalainen et al. 2014). A complete global network would cost billions of dollars, so it remains to be seen where the collection of projects ends up.

Integrated projects are hard to develop because they must satisfy many of the social and political dimensions of infrastructure to be successful. Many infrastructures are domain-focused and then broaden or link to others, but all are embedded in large social systems. There are many shared repositories and discoverability systems that offer search tools and metadata. However, they are scattered across research domains and institutional contexts. Large international enterprises with roots extending back decades, like the Global Earth Observation System of Systems (GEOSS; introduced in Section 2.2.2), are embedded in larger socioeconomic and political systems that differ in how they govern people, distribute tasks and actions, and their underlying governmental principles. None of these actions or responsibilities is neutral. At its heart is the recognition that much human-environment science does not occur in a controlled laboratory setting but instead in a complex and unpredictable reality where standards of repeatable experimentation are no longer possible to implement (Lövbrand et al. 2009). A corollary of this broader and more complicated setting is that scholars must be more aware of, and engaged with, the political, socioeconomic, cultural, and ethical dimensions of research.

Integration is technically, socially, and politically complex. The European Union ENVRIplus project has worked for years to integrate two dozen domain-specific Earth science research infrastructures as a single virtual research community. Its primary contribution is advocating the use of "reference models" to engender interoperability among existing regional and domain-specific infrastructures (Zhao & Hellström 2020). This effort is compelling because it draws together an extensive array of existing infrastructure projects. These include internet hubs and high-speed networks, grid-based computing, and advanced storage that stitch together and support real-world sensor networks on sea, land, and space (Jeffery et al. 2020). This work exemplifies the turn toward big Earth data analytics based on cyberinfrastructure and integration with open standards for interacting with other systems. This approach supports queries, analysis, and visualization in research domains, including marine science, geology, and atmospheric science (Baumann et al. 2016).

The Global Biodiversity Information Facility (GBIF) is another project that illustrates the advantages and disadvantages of extensive infrastructure. The GBIF is a vast repository of biodiversity information with close to a billion records. It draws together a broad range of sources via data standards for manual and automatic information retrieval and collation into an index of millions of species occurrence records. The system includes one of the best-known standards in biological research, the Darwin Core Standard, which specifies the formatting, publishing, and integration of biodiversity data (Wieczorek et al. 2012). The GBIF is especially valuable to researchers in enabling authors and publishers to provide open access their data sets using machine-readable metadata. These metadata feed the creation of hundreds of publications and other products on a range of human-environment topics, including conservation and protected areas, food security, and human well-being. The GBIF is a very successful infrastructure by many measures.

Of course, even a successful and focused infrastructure can be hard to use and maintain for data science. The GBIF supports biodiversity informatics. This work often focuses on estimating the environmental requirements of a given species to establish ecological niches or spatial distribution. While these data are helpful, some experts find that they have the same flaws as other data drawn from multiple sources, including incomplete access to all indexed data. When data are available, they can be inconsistent, incomplete, or wrong. In terms of basic infrastructural dimensions, this infrastructure becomes visible when it breaks down. There is a subsequent need for data quality indicators or uncertainty, preferably those that can be developed in real-time for user-defined regions and time scales. Also needed is better training and guidance for users on accessing and deploying these data in real-world projects (Anderson et al. 2016). There is also a need for information that makes GBIF data sets more helpful in standard analyses. "Scaling up, species traits, population time series, ecological interactions, environmental factors, among others, often need to be integrated with species occurrences to improve the under-standing of ecological process and mechanisms and to test theory" (Poisot et al. 2019, p. 495).

There remain many basic and unmet computing infrastructure needs in dealing with the spatial, temporal, and attribute complexity of human-environment data. For example, there are many large-scale computing frameworks for regional and global Earth observation and analysis (Sudmanns et al. 2019). Many serve remotely sensed imagery and allow some analysis. Google Earth Engine exemplifies this increasingly common approach to some forms of data analysis. It offers powerful computing, an impressive range of operations, and functions that can operate over an extensive data catalog. The system also offers the Earth Engine application program interface, allowing users to use their data with existing data sets. Google Earth Engine works very well for some kinds of tasks. However, there are limits on the size of some kinds of queries, especially those involving large extents, and analyses, particularly those that require computations linking cells in one location to those in another, like classic viewshed or watershed operations (Gorelick et al. 2017). While it offers hundreds of functions, it can be challenging to create complex queries. There is no simple route to getting new functions implemented apart from trying to cobble together new ones from existing ones. Most importantly, there is no guarantee that this product will be continued. Google is a private firm and has a history of shedding dozens of projects and platforms, often with little advance notice.

Divides abound in infrastructure access, limiting some scholarly communities' ability to participate in shared research. Accessing and applying big data requires specialized personnel and infrastructure limited primarily to researchers associated with large companies or universities in high-income, industrialized countries. Similar divides exist in expertise, which tends to be distributed unevenly. For example, geoscientists point to their inability to find the data that they want online, the challenges posed by the heterogeneity of data and metadata, the need for software to use these data, and, more broadly, the sense that funding for long-term cyberinfrastructure funding is insufficient or unstable (Stocks et al. 2019). As explored in Chapter 4, much human-environment research is being carried out by researchers who possess only cursory knowledge about the topics that they purport to study. At the same time, many domain researchers struggle with the core methods of data science.

Public versus private infrastructure is another divide that may limit reach or scope. In some instances, the only people with access to big data are the researchers working for private firms. Most companies will share limited samples but may charge for greater amounts of data or not share them. Research from complete data sets is impossible to verify or replicate absent these data (more in Section 3.4). It is also reasonable to expect researchers to shy away from asking questions that threaten their data supply (boyd & Crawford 2012). While many big data are collected and disseminated by private firms, public agencies are vital in developing critical infrastructure. For example, local and regional governments in many countries capture and share high-quality lidar data. These are used by many agencies for various purposes (e.g., zoning, land planning, emergency response) but are also increasingly used by nongovernmental groups and other publics (Campbell & Salomonson 2010).

Technical challenges compound social and institutional issues when creating human-environment data infrastructure. There is much work in creating or adopting abstract file management systems. For example, the Integrated Rule-Oriented Data System (iRODS) (Xu et al. 2017) offers indexing and search functions for researchers to identify, use, and document subsets of databases. Although there are spatial and temporal issues with these data, scholars derive value from linking across data sets. Linked data should be temporally aligned to measure the presumed causes measured simultaneously or before the effects of interest (Entwisle et al. 2017). There is also exciting work on downstream infrastructures, such as the Open Geospatial Consortium advocating for the Sensor Web Enablement (SWE) initiative. This framework of open standards is designed to allow the producers and users of sensors to connect these devices to the Internet, including "flood gauges, air pollution monitors, stress gauges on bridges, mobile heart monitors, Webcams, airborne and satellite-borne earth imaging devices and countless other sensors and sensor systems" (Botts et al. 2008, p. 2).

Finally, infrastructure is expensive. Large national-scale human-environment data structures and their attendant research can cost vast sums. Even straightforward static sensing systems require a good deal of investment and may not have many other facets of larger infrastructure projects (Mayernik et al. 2013). These systems are inflexible, being fixed in one place and usually having a standard configuration. They are usually limited to measuring specific parameters and not others, as when capturing physical parameters such as soil pH but not many kinds of biological ones, such as biological activity. They also require maintenance, especially in the case of advanced technology such as wireless data collection or sensors needing validation or calibration, and can be challenging to scale up. On the other hand, they are patient and tireless and can collect observations over long timescales. Well-maintained standardized equipment can reduce error and observer bias.

The TeraGrid and XSEDE projects exemplify costly but vital large-scale scientific infrastructure. TeraGrid was founded in 2001. In 2011, it transitioned to the Extreme Science and Engineering Discovery Environment (XSEDE, noted in Section 5.3.1). TeraGrid was one of the first large-scale grid computing infrastructures, drawing on the computing resources from eleven partner institutions and close to US$250 million over a decade from the NSF (Wilkins-Diehr et al. 2008). The infrastructure migrated to XSEDE in the form of a consortium of over a dozen institutions with US$130 million in federal funding (Towns et al. 2014). TeraGrid and XSEDE offer high-performance computing and access to dozens of domain-specific databases. They also provide gateway functionality that integrates the work of technical experts from member institutions to grow computing into new domains or tackle common problems, such as visualization, training, and user engagement. A commitment to ongoing investment in research infrastructure is essential to deal with the seemingly ever-present challenge of creating a "sustainable" computing infrastructure.

Uneven funding regimes affect the development of integrated projects. While there have always been imbalances among fields in their funding, Sawyer (2008) argues that some fields are getting more resources because they traffic in big data. Some fields have data wealth, while others are data-poor, and actual research wealth in the form of funding follows this imbalance. This imbalance is self-perpetuating because research dollars flow to infrastructure in some fields more than others. These enriched fields are better positioned to conduct research and win even more support. There is a danger that many scholars (and funders and government agencies) see big data as a way to supplant other kinds of data entirely. The dream for some scholars is that, for example, ecological monitoring networks would replace the venerated field trip, which is valuable in so many ways to scientific work but tremendously costly and time-consuming (Edwards et al. 2013). As noted in Section 4.3 on small data, other researchers would treat this suggestion with skepticism or even something akin to horror.

Sustainability is central to many conversations around the advantages and disadvantages of research computing capability.

We must eradicate the fallacy that someone else will pay for these capabilities after the initial funding to establish them has expired. Consumers of the services and resources must ultimately pay for them, either directly or via a sponsor. In recent years, there has been a significant shift toward universities supporting a growing share to enable their researchers. In reality, this only shifts the problem rather than identifying much more cost-effective means of providing these capabilities.   *(Towns 2018, p. 44)*

Awardees of large infrastructure grants can try to be as efficient as possible, for instance, by looking to cloud computing (Section 3.3.4). However, it is ultimately incumbent upon government research agencies to continue supporting these projects if they want to see long-term and widespread infrastructure. The growth in the availability of robust computing systems is a significant advance in research computing. It allows individual scholars and small teams to focus on their concrete research questions and leave the management of complex computing environments to others. These shared systems also set the stage for reproducible data science (Section 3.4) and greater interdisciplinarity in data, method, and theory.

6.4 Final Thoughts

Once we get past the starry-eyed embrace of data science or its often justified critiques, there is a middle ground or many potential middle grounds in human-environment research. This thinking sounds Pollyannaish, but we must collaborate to deal with our collective challenges. It can be hard to remain hopeful in the face of what seems to be one unfolding calamity after another – war, famine, disease, disaster – that are usually expressions of long-running human-environment dynamics. Nonetheless, we are making many forms of progress, and scholarly inquiry is part of that effort. Data science is at the center of a broad collection of efforts to build systems and services that support collaborative research on human-environment systems. Data science is also the locus of much interest and training for next-generation scholars and data professionals who will drive future research, ideally in ways that recognize the need for transdisciplinary collaborations among data scientists and domain experts. This shared enterprise should include people working on data science's many significant legal, social, political, and ethical dimensions for human-environment research. Moreover, this work extends to other people and the natural world. We are those people, and that world is us.

References

Abebe, R., Barocas, S., Kleinberg, J. et al. (2020). Roles for Computing in Social Change. In *Proceedings of the 2020 Conference on Fairness, Accountability, and Transparency*, New York: Association for Computing Machinery, pp. 252–260.

Acemoglu, D., & Restrepo, P. (2018). Artificial Intelligence, Automation, and Work. In A. Agrawal, J. Gans, & A. Goldfarb, eds., *The Economics of Artificial Intelligence: An Agenda*, Chicago: University of Chicago Press, pp. 197–236.

Adams, L. G. (2014). Putting together a scientific team: Collaborative science. *Trends in Microbiology*, **22**(9), 483–485.

Agnew, J., Gillespie, T., Gonzalez, J., & Min, B. (2008). Baghdad nights: Evaluating the US military "surge" using nighttime light signatures. *Environment and Planning A*, **40**(10), 2285–2295.

Agrawal, A., Gans, J. S., & Goldfarb, A. (2019). Artificial intelligence: The ambiguous labor market impact of automating prediction. *Journal of Economic Perspectives*, **33**(2), 31–50.

Ahlqvist, O. (2008). Extending post-classification change detection using semantic similarity metrics to overcome class heterogeneity: A study of 1992 and 2001 US National Land Cover Database changes. *Remote Sensing of Environment*, **112**(3), 1226–1241.

Akkineni, V., Aydin, B., Naduvil-Vadukootu, S., & Angryk, R. (2016). Predictive Spatio-Temporal Query Processor on Resilient Distributed Datasets. In *2016 IEEE International Conferences on Big Data and Cloud Computing (BDCloud), Social Computing and Networking (SocialCom), Sustainable Computing and Communications (SustainCom) (BDCloud-SocialCom-SustainCom)*, Santa Clara, CA: Institute of Electrical and Electronics Engineers, pp. 50–58.

Al-Ali, A. R., Zualkernan, I. A., Rashid, M., Gupta, R., & Alikarar, M. (2017). A smart home energy management system using IoT and big data analytics approach. *IEEE Transactions on Consumer Electronics*, **63**(4), 426–434.

Al-Ekabi, C. (2015). The Evolution of Europe's Launcher and Flagship Space Initiatives. In C. Al-Ekabi, ed., *European Autonomy in Space*, Cham: Springer, pp. 1–45.

Alonso, W., & Starr, P. (1987). *The Politics of Numbers*, New York: Russell Sage Foundation.

Altheide, D. L. (2006). Terrorism and the politics of fear. *Cultural Studies ↔ Critical Methodologies*, **6**(4), 415–439.

Amodei, D., Hernandez, D., Sastry, G. et al. (2019). AI and Compute. Accessed January 1, 2020, https://openai.com/blog/ai-and-compute/.

Andersen, R. (2020). The Panopticon Is Already Here. *The Atlantic Monthly*, September. Accessed July 23, 2022, www.theatlantic.com/magazine/archive/2020/.

Anderson, C. (2008). The End of Theory. Wired, June 23, 2008. Accessed July 23, 2022, www.wired.com/2008/06/pb-theory/.

Anderson, G. J., & Lucas, D. D. (2018). Machine learning predictions of a multiresolution climate model ensemble. *Geophysical Research Letters*, **45**(9), 4273–4280.

Anderson, R. P., Araújo, M. B., Guisan, A. et al. (2016). *Final Report of the Task Group on GBIF Data Fitness for Use in Distribution Modelling*, Copenhagen: Global Biodiversity Information Facility.

Anthony, R. E., Ringler, A. T., Wilson, D. C., & Wolin, E. (2018). Do low-cost seismographs perform well enough for your network? An overview of laboratory tests and field observations of the OSOP Raspberry Shake 4D. *Seismological Research Letters*, **90**(1), 219–228.

Antle, J. M., Jones, J. W., & Rosenzweig, C. (2017). Next generation agricultural system models and knowledge products: Synthesis and strategy. *Agricultural Systems*, 155, 179–185.

Aronova, E., Baker, K. S., & Oreskes, N. (2010). Big science and big data in biology: From the International Geophysical Year through the International Biological Program to the Long Term Ecological Research (LTER) Network, 1957–Present. *Historical Studies in the Natural Sciences*, **40**(2), 183–224.

Arora, P. (2016). Bottom of the data pyramid: Big data and the Global South. *International Journal of Communication*, **10**, 1681–1699.

Arribas-Bel, D., Kourtit, K., Nijkamp, P., & Steenbruggen, J. (2015). Cyber cities: Social media as a tool for understanding cities. *Applied Spatial Analysis and Policy*, **8**(3), 231–247.

Arribas-Bel, D., Patino, J. E., & Duque, J. C. (2017). Remote sensing-based measurement of living environment deprivation: Improving classical approaches with machine learning. *PLOS ONE*, **12**(5), e0176684.

Arts, K., van der Wal, R., & Adams, W. M. (2015). Digital technology and the conservation of nature. *Ambio*, **44**(4), 661–673.

Ash, J., Kitchin, R., & Leszczynski, A. (2016). Digital turn, digital geographies? *Progress in Human Geography*, **42**(1), 25–43.

Ashton, K. (2009). That "Internet of things" thing. *RFID Journal*, **22**(7), 97–114.

Athey, S. (2019). The Impact of Machine Learning on Economics. In A. Goldfarb, J. Gans, & A. Agrawal, eds., *The Economics of Artificial Intelligence: An Agenda*, Chicago: University of Chicago Press, pp. 507–547.

Athey, S., & Imbens, G. W. (2019). Machine learning methods that economists should know about. *Annual Review of Economics*, **11**, 685–725.

August, T., Harvey, M., Lightfoot, P. et al. (2015). Emerging technologies for biological recording. *Biological Journal of the Linnean Society*, **115**(3), 731–749.

Azzari, G., & Lobell, D. B. (2017). Landsat-based classification in the cloud: An opportunity for a paradigm shift in land cover monitoring. *Remote Sensing of Environment*, **202**, 64–74.

Bagrow, J. P., Wang, D., & Barabási, A.-L. (2011). Collective response of human populations to large-scale emergencies. *PLOS ONE*, **6**(3), e17680.

Baker, K. S., Benson, B. J., Henshaw, D. L. et al. (2000). Evolution of a multisite network information system: The LTER Information Management Paradigm. *BioScience*, **50**(11), 963–978.

Balk, D., Storeygard, A., Levy, M. et al. (2005). Child hunger in the developing world: An analysis of environmental and social correlates. *Food Policy*, **30**(5–6), 584–611.

Bantwal Rao, M., Jongerden, J., Lemmens, P., & Ruivenkamp, G. (2015). Technological mediation and power: Postphenomenology, critical theory, and autonomist Marxism. *Philosophy & Technology*, **28**(3), 449–474.

Baraniuk, R. G. (2011). More is less: Signal processing and the data deluge. *Science*, **331** (6081), 717–719.

Baro, E., Degoul, S., Beuscart, R., & Chazard, E. (2015). Toward a literature-driven definition of big data in healthcare. *BioMed Research International*, **2015**, 639021.

Barocas, S., & Selbst, A. D. (2016). Big data's disparate impact. *California Law Review*, **104**, 671–732.

Barth, K.-H. (2003). The politics of seismology: Nuclear testing, arms control, and the transformation of a discipline. *Social Studies of Science*, **33**(5), 743–781.

Bartholome, E., & Belward, A. S. (2005). GLC2000: A new approach to global land cover mapping from Earth observation data. *International Journal of Remote Sensing*, **26**(9), 1959–1977.

Bassil, K. L., Sanborn, M., Lopez, R., & Orris, P. (2015). Integrating environmental and human health databases in the Great Lakes basin: Themes, challenges and future directions. *International Journal of Environmental Research and Public Health*, **12**(4), 3600–3614.

Bates, J. (2014). The strategic importance of information policy for the contemporary neoliberal state: The case of Open Government Data in the United Kingdom. *Government Information Quarterly*, **31**(3), 388–395.

Bates, J. (2017). Big Data, Open Data and the Climate Risk Market. In B. Brevini & G. Murdock, eds., *Carbon Capitalism and Communication: Confronting Climate Crisis*, Cham: Springer, pp. 83–93.

Bates, J., & Goodale, P. (2017). Making data flow for the climate risk market. *Television & New Media*, **18**(8), 753–768.

Batty, M. (2013). *The New Science of Cities*, Cambridge, MA: MIT Press.

Batty, M., Axhausen, K. W., Giannotti, F. et al. (2012). Smart cities of the future. *The European Physical Journal Special Topics*, **214**(1), 481–518.

Baumann, P., Mazzetti, P., Ungar, J. et al. (2016). Big data analytics for Earth sciences: The EarthServer approach. *International Journal of Digital Earth*, **9**(1), 3–29.

Ben-Nun, T., & Hoefler, T. (2019). Demystifying parallel and distributed deep learning: An in-depth concurrency analysis. *ACM Computing Surveys (CSUR)*, **52**(4), 1–43.

Bengtsson, L., Gaudart, J., Lu, X. et al. (2015). Using mobile phone data to predict the spatial spread of cholera. *Scientific Reports*, **5**(1), 1–5.

Bernards, N. (2022). Colonial financial infrastructures and Kenya's uneven fintech boom. *Antipode*, **54**(3), 708–728.

Berry, D. M. (2011). The computational turn: Thinking about the digital humanities. *Culture Machine*, **12**, 1–22.

Bettencourt, L., & West, G. (2010). A unified theory of urban living. *Nature*, **467**(7318), 912–913.

Bhandari Neupane, J., Neupane, R. P., Luo, Y. et al. (2019). Characterization of leptazolines A–D, polar oxazolines from the cyanobacterium leptolyngbya sp., reveals a glitch with the "Willoughby–Hoye" scripts for calculating NMR chemical shifts. *Organic Letters*, **21**(20), 8449–8453.

Bibri, S. E., & Krogstie, J. (2017). Smart sustainable cities of the future: An extensive interdisciplinary literature review. *Sustainable Cities and Society*, **31**, 183–212.

Birenboim, A., & Shoval, N. (2016). Mobility research in the age of the smartphone. *Annals of the American Association of Geographers*, **106**(2), 283–291.

Bishop, B. W., & Hank, C. (2018). Earth Science Data Management: Mapping Actual Tasks to Conceptual Actions in the Curation Lifecycle Model. In G. Chowdhury, J. McLeod, V. Gillet, & P. Willett, eds., *Transforming Digital Worlds, iConference 2018*, Cham: Springer, pp. 598–608.

Blaikie, P., Cannon, T., Davis, I., & Wisner, B. (1994). *At Risk: Natural Hazards, People's Vulnerability, and Disasters*, New York: Routledge.

Blond, N., Boersma, K. F., Eskes, H. J. et al. (2007). Intercomparison of SCIAMACHY nitrogen dioxide observations, in situ measurements and air quality modeling results over Western Europe. *Journal of Geophysical Research: Atmospheres*, **112**(10), D10311.

Blumenstock, J. E., & Eagle, N. (2012). Divided we call: Disparities in access and use of mobile phones in Rwanda. *Information Technologies & International Development*, **8** (2), 1–16.

Boakes, E. H., Gliozzo, G., Seymour, V. et al. (2016). Patterns of contribution to citizen science biodiversity projects increase understanding of volunteers' recording behaviour. *Scientific Reports*, **6**(1), 33051.

Boeckhout, M., Zielhuis, G. A., & Bredenoord, A. L. (2018). The FAIR guiding principles for data stewardship: Fair enough? *European Journal of Human Genetics*, **26**(7), 931–936.

Boellstorff, T., Helmreich, S., Jones, G. M. et al. (2016). For whom the ontology turns: Theorizing the digital real. *Current Anthropology*, **57**(4), 387–407.

Bollacker, K. D. (2010). Computing science: Avoiding a digital dark age. *American Scientist*, **98**(2), 106–110.

Bolstad, P., & Manson, S. M. (2022). *GIS Fundamentals: A First Text on Geographic Information Systems*, White Bear Lake, MN: Eider Press.

Bongirwar, V. (2020). Stochastic event set generation for tropical cyclone using machine-learning approach guided by environmental data. *International Journal of Climatology*, **40**(15), 6265–6281.

Bordogna, G., Carrara, P., Criscuolo, L., Pepe, M., & Rampini, A. (2016). On predicting and improving the quality of Volunteer Geographic Information projects. *International Journal of Digital Earth*, **9**(2), 134–155.

Borgman, C. L. (2019). The lives and after lives of data. *Harvard Data Science Review*, **1** (1), https://doi.org/10.1162/99608f92.9a36bdb6.

Borgman, C. L., Wallis, J. C., & Mayernik, M. S. (2012). Who's got the data? Interdependencies in science and technology collaborations. *Computer Supported Cooperative Work (CSCW)*, **21**(6), 485–523.

Boschetti, A., & Massaron, L. (2018). *Python Data Science Essentials: A Practitioner's Guide Covering Essential Data Science Principles, Tools, and Techniques*, Birmingham: Packt Publishing Ltd.

Botts, M. , Percivall, G. , Reed, C. , & Davidson, J. (2008). OGC Sensor Web Enablement: Overview and High Level Architecture. In S. Nittel, A. Labrinidis, & A. Stefanidis, eds., *International Conference on GeoSensor Networks*, Heidelberg: Springer, pp. 175–190.

Boudon, R. (1991). What middle-range theories are. *Contemporary Sociology*, **20**(4), 519–522.

Boulos, M. N. K., Resch, B., Crowley, D. N. et al. (2011). Crowdsourcing, citizen sensing and sensor web technologies for public and environmental health surveillance and crisis management: Trends, OGC standards and application examples. *International Journal of Health Geographics*, **10**(1), 67.

Bouveyron, C., Celeux, G., Murphy, T. B., & Raftery, A. E. (2019). *Model-Based Clustering and Classification for Data Science: With Applications in R*, Cambridge: Cambridge University Press.

boyd, D., & Crawford, K. (2012). Critical questions for big data: Provocations for a cultural, technological, and scholarly phenomenon. *Information Communication and Society*, **15** (5), 662–679.

Brackstone, G. (1999). Managing data quality in a statistical agency. *Survey Methodology*, **25**(2), 139–150.

Brady, H. E. (2019). The challenge of big data and data science. *Annual Review of Political Science*, **22**(1), 297–323.

Braman, S. (2006). *Change of State: Information, Policy, and Power*, Cambridge, MA: MIT Press.

Braun, B. (2004). Querying posthumanisms. *Geoforum*, 3(35), 269–273.

Brayne, S. (2017). Big data surveillance: The case of policing. *American Sociological Review*, **82**(5), 977–1008.

British Academy. (2012). *Society Counts: Quantitative Skills in the Social Sciences and Humanities*, London: British Academy.

Brodie, M. L. (1984). On the Development of Data Models. In M. L. Brodie, J. Mylopoulos, & J. W. Schmidt, eds., *On Conceptual Modelling: Perspectives from Artificial Intelligence, Databases, and Programming Languages*, New York: Springer, pp. 19–47.

Brynjolfsson, E., & McAfee, A. (2014). *The Second Machine Age: Work, Progress, and Prosperity in a Time of Brilliant Technologies*, New York: WW Norton & Company.

Buckland, M. K. (1991). Information as thing. *Journal of the American Society for Information Science*, **42**(5), 351–360.

Buda, M., Maki, A., & Mazurowski, M. A. (2018). A systematic study of the class imbalance problem in convolutional neural networks. *Neural Networks*, **106**, 249–259.

Buenemann, M., Martius, C., Jones, J. W. et al. (2011). Integrative geospatial approaches for the comprehensive monitoring and assessment of land management sustainability: Rationale, potentials, and characteristics. *Land Degradation & Development*, **22**(2), 226–239.

Buolamwini, J., & Gebru, T. (2018). Gender Shades: Intersectional Accuracy Disparities in Commercial Gender Classification. In *Conference on Fairness, Accountability and Transparency*, Maastricht: ML Research Press, pp. 77–91.

Burns, R. (2015). Rethinking big data in digital humanitarianism: Practices, epistemologies, and social relations. *GeoJournal*, **80**(4), 477–490.

Burton, A. C., Neilson, E., Moreira, D. et al. (2015). Wildlife camera trapping: A review and recommendations for linking surveys to ecological processes. *Journal of Applied Ecology*, **52**(3), 675–685.

Busch, J., & Ferretti-Gallon, K. (2017). What drives deforestation and what stops it? A meta-analysis. *Review of Environmental Economics and Policy*, **11**(1), 3–23.

Butler, R., Lay, T., Creager, K. et al. (2004). The global seismographic network surpasses its design goal. *Eos, Transactions American Geophysical Union*, **85**(23), 225–229.

Buzzelli, M. (2020). Modifiable Areal Unit Problem. In A. Kobayashi, ed., *International Encyclopedia of Human Geography*, Amsterdam: Elsevier, pp. 169–173.

Caldwell, P. M., Bretherton, C. S., Zelinka, M. D. et al. (2014). Statistical significance of climate sensitivity predictors obtained by data mining. *Geophysical Research Letters*, **41**(5), 1803–1808.

Campbell, A. T., Eisenman, S. B., Lane, N. D. et al. (2008). The rise of people-centric sensing. *IEEE Internet Computing*, **12**(4), 12–21.

Campbell, J. B., & Salomonson, V. V. (2010). Remote Sensing: A Look to the Future. In J. D. Bossler, J. R. Jensen, R. B. McMaster, & C. Rizos, eds., *Manual of Geospatial Science and Technology*, Boca Raton, FL: CRC Press, pp. 487–509.

Carolan, M. (2017). Publicising food: Big data, precision agriculture, and co-experimental techniques of addition. *Sociologia Ruralis*, **57**(2), 135–154.

Carson, C. (1998). Fostering the fundamental principles of official statistics. *Statistical Journal of the UN Economic Commission for Europe*, **15**(3–4), 213–220.

Casana, J. (2020). Global-scale archaeological prospection using CORONA satellite imagery: Automated, crowd-sourced, and expert-led approaches. *Journal of Field Archaeology*, **45**(suppl. 1), S89–S100.

Cass, S. (2019). Taking AI to the edge: Google's TPU now comes in a maker-friendly package. *IEEE Spectrum*, **56**(5), 16–17.

Castell, N., Dauge, F. R., Schneider, P. et al. (2017). Can commercial low-cost sensor platforms contribute to air quality monitoring and exposure estimates? *Environment International*, **99**, 293–302.

Cavallo, A., & Rigobon, R. (2016). The Billion Prices Project: Using online prices for measurement and research. *Journal of Economic Perspectives*, **30**(2), 151–178.

Chang, E. K. M., & Guo, Y. (2007). Is the number of North Atlantic tropical cyclones significantly underestimated prior to the availability of satellite observations? *Geophysical Research Letters*, **34**(14), 547–572.

Chang, W. L., & Grady, N. (2015). *NIST Big Data Interoperability Framework: Volume 1, Big Data Definitions*, Gaithersburg, MD: National Institute of Standards and Technology.

Chapman, L., Bell, C., & Bell, S. (2017). Can the crowdsourcing data paradigm take atmospheric science to a new level? A case study of the urban heat island of London quantified using Netatmo weather stations. *International Journal of Climatology*, **37**(9), 3597–3605.

Charalabidis, Y., Zuiderwijk, A., Alexopoulos, C. et al. (2018). The Open Data Landscape. In Y. Charalabidis, A. Zuiderwijk, C. Alexopoulos et al., eds., *The World of Open Data: Concepts, Methods, Tools and Experiences*, Cham: Springer, pp. 1–9.

Checkland, P., & Holwell, S. (2006). Data, Capta, Information and Knowledge. In M. Hinton, ed., *Introducing Information Management: The Business Approach*, London: Routledge, pp. 47–55.

Chen, F., & Neill, D. B. (2015). Human rights event detection from heterogeneous social media graphs. *Big Data*, **3**(1), 34–40.

Chen, J., Chen, J., Liao, A. et al. (2015). Global land cover mapping at 30m resolution: A POK-based operational approach. *ISPRS Journal of Photogrammetry and Remote Sensing*, **103**, 7–27.

Chen, M., Mao, S., & Liu, Y. (2014). Big data: A survey. *Mobile Networks and Applications*, **19**(2), 171–209.

Chen, X.-W., & Lin, X. (2014). Big data deep learning: Challenges and perspectives. *IEEE Access*, **2**, 514–525.

Chen, Z., Pan, H., Liu, C., & Jiang, Z. (2018). Agricultural Remote Sensing and Data Science in China. In F. A. Batarseh & R. Yang, eds., *Federal Data Science: Transforming Government and Agricultural Policy Using Artificial Intelligence*, Cambridge, MA: Academic Press, pp. 95–108.

Chen, Z., & Liu, B. (2016). Lifelong machine learning. *Synthesis Lectures on Artificial Intelligence and Machine Learning*, **10**(3), 1–145.

Chipman, H. A., & Joseph, V. R. (2016). A conversation with Jeff Wu. *Statistical Science*, **31**(4), 624–636.

Christie, M. (2004). Data collection and the ozone hole. *History of Meteorology*, **1**, 99–105.

Christin, D., Reinhardt, A., Kanhere, S. S., & Hollick, M. (2011). A survey on privacy in mobile participatory sensing applications. *Journal of Systems and Software*, **84**(11), 1928–1946.

Cinnamon, J. (2022). On data cultures and the prehistories of smart urbanism in "Africa's Digital City." *Urban Geography*, **43**(forthcoming), https://doi.org/10.1080/02723638 .2022.2049096.

Cirac-Claveras, G. (2018). The weather privateers: Meteorology and commercial satellite data. *Information & Culture*, **53**(3/4), 271–302.

Clark, W. R., & Golder, M. (2015). Big data, causal inference, and formal theory: Contradictory trends in political science? *PS: Political Science & Politics*, **48**(1), 65–70.

Clerx, M., Cooling, M. T., Cooper, J. et al. (2020). CellML 2.0. *Journal of Integrative Bioinformatics*, **17**(2–3), 20200021.

Cleveland, W. S. (2001). Data science: An action plan for expanding the technical areas of the field of statistics. *International Statistical Review*, **69**(1), 21–26.

Combi, C., & Pozzi, G. (2019). Clinical information systems and artificial intelligence: Recent research trends. *Yearbook of Medical Informatics*, **28**(1), 83–94.

Connors, J. P., Lei, S., & Kelly, M. (2012). Citizen science in the age of neogeography: Utilizing volunteered geographic information for environmental monitoring. *Annals of the Association of American Geographers*, **102**(6), 1267–1289.

Cooper, C. B., Hochachka, W. M., & Dhondt, A. A. (2012). The Opportunities and Challenges of Citizen Science as a Tool for Ecological Research. In J. L. Dickinson & R. E. Bonney, eds., *Citizen Science: Public Participation in Environmental Research*, Ithaca, NY: Cornell University Press, pp. 99–113.

Cooper, M. (2010). Turbulent worlds. *Theory, Culture & Society*, **27**(2–3), 167–190.

Cooper, N., Hsing, P.-Y., Croucher, M. et al. (2017). *A Guide to Reproducible Code in Ecology and Evolution*, London: British Ecological Society.

Costanza, R., D'Arge, R., de Groot, R. et al. (1997). The value of the world's ecosystem services and natural capital. *Nature*, **387**(6630), 253–260.

Couclelis, H. (2003). The certainty of uncertainty: GIS and the limits of geographic knowledge. *Transactions in GIS*, **7**(2), 165–175.

Coveney, P. V., Dougherty, E. R., & Highfield, R. R. (2016). Big data need big theory too. *Philosophical Transactions of the Royal Society A: Mathematical, Physical and Engineering Sciences*, **374**(2080), 20160153.

Cowan, J. D. (1995). *Fault and Failure Tolerance*, Chicago: The University of Chicago.

Crabtree, A., Tolmie, P., & Knight, W. (2017). Repacking "privacy" for a networked world. *Computer Supported Cooperative Work (CSCW)*, **26**(4), 453–488.

Craglia, M., de Bie, K., Jackson, D. et al. (2012). Digital earth 2020: Towards the vision for the next decade. *International Journal of Digital Earth*, **5**(1), 4–21.

Crall, A. W., Newman, G. J., Stohlgren, T. J. et al. (2011). Assessing citizen science data quality: An invasive species case study. *Conservation Letters*, **4**(6), 433–442.

Crawford, C. J., Manson, S. M., Bauer, M. E., & Hall, D. K. (2013). Multitemporal snow cover mapping in mountainous terrain for Landsat climate data record development. *Remote Sensing of Environment*, **135**, 224–233.

Crawford, K., & Finn, M. (2015). The limits of crisis data: Analytical and ethical challenges of using social and mobile data to understand disasters. *GeoJournal*, **80**(4), 491–502.

Crawl, D., Singh, A., & Altintas, I. (2016). Kepler webview: A lightweight, portable framework for constructing real-time web interfaces of scientific workflows. *Procedia Computer Science*, **80**, 673–679.

Crews, K. A., & Walsh, S. J. (2009). Remote Sensing and the Social Sciences. In T. A. Warner, M. D. Nellis, & G. M. Foody, eds., *The SAGE Handbook of Remote Sensing*, London: SAGE Publications, pp. 437–445.

Crosas, M., King, G., Honaker, J., & Sweeney, L. (2015). Automating open science for big data. *The Annals of the American Academy of Political and Social Science*, **659**(1), 260–273.

Crutcher, M., & Zook, M. (2009). Placemarks and waterlines: Racialized cyberscapes in post-Katrina Google Earth. *Geoforum*, **40**(4), 523–534.

Cuff, D., Hansen, M., & Kang, J. (2008). Urban sensing: Out of the woods. *Communications of the ACM*, **51**(3), 24–33.

Cutcher-Gershenfeld, J., Baker, K. S., Berente, N. et al. (2016). Build it, but will they come? A geoscience cyberinfrastructure baseline analysis. *Data Science Journal*, **15**, 8.

Cuzzocrea, A., Song, I.-Y., & Davis, K. C. (2011). Analytics over Large-scale Multidimensional Data: The Big Data Revolution! In *Proceedings of the ACM 14th International Workshop on Data Warehousing and OLAP*, Glasgow: Association for Computing Machinery, pp. 101–104.

d'Alessandro, B., O'Neil, C., & LaGatta, T. (2017). Conscientious classification: A data scientist's guide to discrimination-aware classification. *Big Data*, **5**(2), 120–134.

Datta, A. (2015). New urban utopias of postcolonial India: "Entrepreneurial urbanization" in Dholera smart city, Gujarat. *Dialogues in Human Geography*, **5**(1), 3–22.

Davies, J., Studer, R., & Warren, P. (2006). *Semantic Web Technologies: Trends and Research in Ontology-Based Systems*, Hoboken, NJ: John Wiley & Sons.

De-Arteaga, M., Herlands, W., Neill, D. B., & Dubrawski, A. (2018). Machine learning for the developing world. *ACM Transactions on Management Information Systems (TMIS)*, **9**(2), 9.

de Bruijn, J. A., de Moel, H., Jongman, B. et al. (2019). A global database of historic and real-time flood events based on social media. *Scientific Data*, **6**(1), 311.

de Donno, M., Tange, K., & Dragoni, N. (2019). Foundations and evolution of modern computing paradigms: Cloud, IoT, Edge, and Fog. *IEEE Access*, **7**, 150936–150948.

de Longueville, B., Annoni, A., Schade, S., Ostlaender, N., & Whitmore, C. (2010). Digital Earth's nervous system for crisis events: Real-time sensor web enablement of volunteered geographic information. *International Journal of Digital Earth*, **3**(3), 242–259.

de Nazelle, A., Seto, E., Donaire-Gonzalez, D. et al. (2013). Improving estimates of air pollution exposure through ubiquitous sensing technologies. *Environmental Pollution*, **176**, 92–99.

de Vries, K. (2010). Identity, profiling algorithms and a world of ambient intelligence. *Ethics and Information Technology*, **12**(1), 71–85.

Dean, J., & Ghemawat, S. (2008). MapReduce: Simplified data processing on large clusters. *Communications of the ACM*, **51**(1), 107–113.

DeFries, R. (2008). Terrestrial vegetation in the coupled human-earth system: Contributions of remote sensing. *Annual Review of Environment and Resources*, **33**(1), 369–390.

Degrossi, L. C., Porto de Albuquerque, J., Santos Rocha, R. dos, & Zipf, A. (2018). A taxonomy of quality assessment methods for volunteered and crowdsourced geographic information. *Transactions in GIS*, **22**(2), 542–560.

D'Ignazio, C. , & Klein, L. (2020). *Data Feminism*. Cambridge, MA: MIT Press.

Del Vicario, M., Bessi, A., Zollo, F. et al. (2016). The spreading of misinformation online. *Proceedings of the National Academy of Sciences*, **113**(3), 554–559.

DeMasi, O., Paxton, A., & Koy, K. (2020). Ad hoc efforts for advancing data science education. *PLOS Computational Biology*, **16**(5), e1007695.

Determann, L., Ruan, Z. J., Gao, T., & Tam, J. (2021). China's draft Personal Information Protection Law. *Journal of Data Protection & Privacy*, **4**(3), 235–259.

Devarajan, S. (2013). Africa's statistical tragedy. *Review of Income and Wealth*, **59**(S1), S9–S15.

Deville, P., Linard, C., Martin, S. et al. (2014). Dynamic population mapping using mobile phone data. *Proceedings of the National Academy of Sciences*, **111**(45), 15888–15893.

di Modica, G., & Tomarchio, O. (2022). A hierarchical hadoop framework to process geo-distributed big data. *Big Data and Cognitive Computing*, **6**(1), 5.

Dias, M. B., & Brewer, E. (2009). How computer science serves the developing world. *Communications of the ACM*, **52**(6), 74–80.

Diebold, F. X. (2012). *A Personal Perspective on the Origin(s) and Development of "Big Data": The Phenomenon, the Term, and the Discipline*, second version, Philadelphia: Penn Institute for Economic Research (PIER).

Dijkgraaf, R. (2021). The Uselessness of Useful Knowledge. Quanta, October 20. Accessed August 11, 2022, www.quantamagazine.org/science-has-entered-a-new-era-of-alchemy-good-20211020/.

Ditmer, M. A., Vincent, J. B., Werden, L. K. et al. (2015). Bears show a physiological but limited behavioral response to unmanned aerial vehicles. *Current Biology*, **25**(17), 2278–2283.

Do, H. X., Westra, S., Leonard, M., & Gudmundsson, L. (2020). Global-scale prediction of flood timing using atmospheric reanalysis. *Water Resources Research*, **56**(1), e2019WR024945.

Dong, X., Yu, Z., Cao, W., Shi, Y., & Ma, Q. (2020). A survey on ensemble learning. *Frontiers of Computer Science*, **14**(2), 241–258.

Donoho, D. (2017). 50 years of data science. *Journal of Computational and Graphical Statistics*, **26**(4), 745–766.

Dunnette, M. D. (1966). Fads, fashions, and folderol in psychology. *American Psychologist*, **21**(4), 343–352.

Duporge, I., Isupova, O., Reece, S., Macdonald, D. W., & Wang, T. (2020). Using very-high-resolution satellite imagery and deep learning to detect and count African elephants in heterogeneous landscapes. *Remote Sensing in Ecology and Conservation*, **7**, 369–381.

Dusek, V. (2006). *Philosophy of Technology: An Introduction*, Oxford: Blackwell Publishing Ltd.

Edwards, P. N., Jackson, S. J., Chalmers, M. K. et al. (2013). *Knowledge Infrastructures: Intellectual Frameworks and Research Challenges*, Ann Arbor, MI: Deep Blue.

Ehrlich, P. R., Kareiva, P. M., & Daily, G. C. (2012). Securing natural capital and expanding equity to rescale civilization. *Nature*, **486**(7401), 68–73.

Elahi, S. (2009). Privacy and consent in the digital era. *Information Security Technical Report*, **14**(3), 113–118.

Eldawy, A., & Mokbel, M. F. (2016). The era of big spatial data: A survey. *Foundations and Trends in Databases*, **6**(3–4), 305–316.

Eldawy, A., Mokbel, M. F., & Jonathan, C. (2016). HadoopViz: A MapReduce Framework for Extensible Visualization of Big Spatial Data. In *2016 IEEE 32nd International Conference on Data Engineering (ICDE)*, Santa Clara, CA: Institute of Electrical and Electronics Engineers, pp. 601–612.

Elfenbein, A. (2020). *The Gist of Reading*, Stanford, CA: Stanford University Press.

Elliott, K. C., Cheruvelil, K. S., Montgomery, G. M., & Soranno, P. A. (2016). Conceptions of good science in our data-rich world. *BioScience*, **66**(10), 880–889.

Entwisle, B., Hofferth, S. L., & Moran, E. F. (2017). Quilting a time-place mosaic: Concluding remarks. *The Annals of the American Academy of Political and Social Science*, **669**(1), 190–198.

Esch, T., Heldens, W., Hirner, A. et al. (2017). Breaking new ground in mapping human settlements from space: The Global Urban Footprint. *ISPRS Journal of Photogrammetry and Remote Sensing*, **134**, 30–42.

Espey, J. (2019). Sustainable development will falter without data. *Nature*, **571**(7765), 299–300.

Essawy, B. T., Goodall, J. L., Xu, H., & Gil, Y. (2017). Evaluation of the OntoSoft ontology for describing metadata for legacy hydrologic modeling software. *Environmental Modelling & Software*, **92**, 317–329.

Essl, A., Ortner, A., Haas, R., & Hettegger, P. (2017). Machine Learning Analysis for a Flexibility Energy Approach towards Renewable Energy Integration with Dynamic Forecasting of Electricity Balancing Power. In *14th International Conference on the European Energy Market (EEM)*, Santa Clara, CA: Institute of Electrical and Electronics Engineers, https://doi.org/10.1109/EEM.2017.7981877.

Fadiya, S. O., Saydam, S., & Zira, V. V. (2014). Advancing big data for humanitarian needs. *Procedia Engineering*, **78**, 88–95.

Faghmous, J. H., & Kumar, V. (2014a). A big data guide to understanding climate change: The case for theory-guided data science. *Big Data*, **2**(3), 155–163.

Faghmous, J. H., & Kumar, V. (2014b). Spatio-temporal Data Mining for Climate Data: Advances, Challenges, and Opportunities. In *Data Mining and Knowledge Discovery for Big Data*, New York: Springer, pp. 83–116.

Fan, J., Han, F., & Liu, H. (2014). Challenges of big data analysis. *National Science Review*, **1**(2), 293–314.

Fan, J., Li, R., Zhang, C.-H., & Zou, H. (2020). *Statistical Foundations of Data Science*, Boca Raton, FL: CRC press.

Farnes, J., Mort, B., Dulwich, F., Salvini, S., & Armour, W. (2018). Science pipelines for the square kilometre array. *Galaxies*, **6**(4), 120.

Favaretto, M., de Clercq, E., & Elger, B. S. (2019). Big data and discrimination: Perils, promises and solutions. A systematic review. *Journal of Big Data*, **6**(1), 12.

Feenberg, A. (2002). *Questioning Technology*, New York: Routledge.

Feenberg, A. (2009). What Is Philosophy of Technology? In A. Jones & M. J. de Vries, eds., *International Handbook of Research and Development in Technology Education*, Leiden: Brill | Sense, pp. 159–166.

Feenberg, A. (2017). *Technosystem*, Cambridge, MA: Harvard University Press.

Filho, C. R. D. S., Zullo Jr., J., & Elvidge, C. (2004). Brazil's 2001 energy crisis monitored from space. *International Journal of Remote Sensing*, **25**(12), 2475–2482.

Fiske, S. T., & Hauser, R. M. (2014). Protecting human research participants in the age of big data. *Proceedings of the National Academy of Sciences*, **111**(38), 13675–13676.

Fleming, L., Haines, A., Golding, B. et al. (2014). Data mashups: Potential contribution to decision support on climate change and health. *International Journal of Environmental Research and Public Health*, **11**(2), 1725–1746.

Fleming, L., Tempini, N., Gordon-Brown, H. et al. (2017). Big Data in Environment and Human Health. In *Oxford Research Encyclopedia of Environmental Science*, Oxford: Oxford University Press, https://doi.org/10.1093/acrefore/9780199389414.013.541.

Fogel, D. B. (1994). An introduction to simulated evolutionary optimization. *IEEE Transactions on Neural Networks*, **1**(1), 3–14.

Fohringer, J., Dransch, D., Kreibich, H., & Schröter, K. (2015). Social media as an information source for rapid flood inundation mapping. *Natural Hazards and Earth System Sciences*, **15**(12), 2725–2738.

Fox, G., & Chang, W. L. (2018). *NIST Big Data Interoperability Framework: Volume 3, Use Cases and General Requirements, Version 2*, Gaithersburg, MD: National Institute of Standards and Technology.

Fox, J., Rindfuss, R. R., Walsh, S. J., & Mishra, V. (2003). *People and the Environment: Approaches for Linking Household and Community Surveys to Remote Sensing and GIS*, Boston: Kluwer Academic Publishers.

Francis, J. G., & Francis, L. P. (2014). Privacy, confidentiality, and justice. *Journal of Social Philosophy*, **45**(3), 408–431.

Franklin, R. S., Delmelle, E. C., Andris, C. et al. (2022). Making space in geographical analysis. *Geographical Analysis*, **54**(forthcoming), https://doi.org/10.1111/gean.12325.

Fraser, A. (2019). Land grab/data grab: Precision agriculture and its new horizons. *The Journal of Peasant Studies*, **46**(5), 893–912.

Frew, J., & Dozier, J. (1997). Data management for Earth system science. *SIGMOD Record*, **26**(1), 27–31.

Friendly, M. (2007). A.-M. Guerry's "Moral Statistics of France": Challenges for multivariable spatial analysis. *Statistical Science*, **22**(3), 368–399.

Fritz, S., & Lee, L. (2005). Comparison of land cover maps using fuzzy agreement. *International Journal of Geographical Information Science*, **19**(7), 787–807.

Future Earth. (2014). *Future Earth Strategic Research Agenda 2014*, Paris: International Council for Science.

Future Earth. (2019). *Annual Report 2017–2018*, Paris: International Council for Science.

Gabrys, J., Pritchard, H., & Barratt, B. (2016). Just good enough data: Figuring data citizenships through air pollution sensing and data stories. *Big Data & Society*, **3**(2), https://doi.org/10.1177/2053951716679677.

Gahegan, M. (2020). Fourth paradigm GIScience? Prospects for automated discovery and explanation from data. *International Journal of Geographical Information Science*, **34** (1), 1–21.

Gallego, J., Rivero, G., & Martínez, J. (2021). Preventing rather than punishing: An early warning model of malfeasance in public procurement. *International Journal of Forecasting*, **37**(1), 360–377.

Gandomi, A., & Haider, M. (2015). Beyond the hype: Big data concepts, methods, and analytics. *International Journal of Information Management*, **35**(2), 137–144.

Ganguly, A. R., Kodra, E. A., Agrawal, A. et al. (2014). Toward enhanced understanding and projections of climate extremes using physics-guided data mining techniques. *Nonlinear Processes in Geophysics*, **21**(4), 777–795.

Gantz, J., & Reinsel, D. (2010). *The Digital Universe Decade – Are You Ready?*, Needham, MA: International Data Corporation (sponsored by EMC).

Gavish, M., & Donoho, D. (2012). Three dream applications of verifiable computational results. *Computing in Science & Engineering*, **14**(4), 26–31.

Gebru, T., Morgenstern, J., Vecchione, B. et al. (2021). Datasheets for datasets. *Communications of the ACM*, **64**(12), 86–92.

Gelaro, R., McCarty, W., Suárez, M. J. et al. (2017). The Modern-Era Retrospective Analysis for Research and Applications, version 2 (MERRA-2). *Journal of Climate*, **30**(14), 5419–5454.

Gentine, P., Pritchard, M., Rasp, S., Reinaudi, G., & Yacalis, G. (2018). Could machine learning break the convection parameterization deadlock? *Geophysical Research Letters*, **45**(11), 5742–5751.

Ghermandi, A. (2018). Integrating social media analysis and revealed preference methods to value the recreation services of ecologically engineered wetlands. *Ecosystem Services*, **31**, 351–357.

Ghermandi, A., & Sinclair, M. (2019). Passive crowdsourcing of social media in environmental research: A systematic map. *Global Environmental Change*, **55**, 36–47.

Gibreel, O., & Hong, A. (2017). A holistic analysis approach to social, technical, and socio-technical aspect of e-government development. *Sustainability*, **9**(12), 2181.

Gil, Y., Garijo, D., Mishra, S., & Ratnakar, V. (2016). OntoSoft: A Distributed Semantic Registry for Scientific Software. In *2016 IEEE 12th International Conference on e-Science (e-Science)*, Santa Clara, CA: Institute of Electrical and Electronics Engineers, pp. 331–336.

Giorgi, F. (2019). Thirty years of regional climate modeling: Where are we and where are we going next? *Journal of Geophysical Research: Atmospheres*, **124**(11), 5696–5723.

Glavovic, B. C., Smith, T. F., & White, I. (2021). The tragedy of climate change science. *Climate and Development*, **14**(forthcoming), https://doi.org/10.1080/17565529.2021.2008855.

Global Earth Observation System of Systems (GEOSS). (2005). *10-Year Implementation Plan Reference Document*, Noordwijk: ESA Publications Division. Accessed August 11, 2022, https://earthobservations.org/documents/10-YearPlanReferenceDocument.pdf.

Gogolenko, S., Groen, D., Suleimenova, D. et al. (2020). Towards Accurate Simulation of Global Challenges on Data Centers Infrastructures via Coupling of Models and Data Sources. In V. V Krzhizhanovskaya, G. Závodszky, M. H. Lees et al., eds., *Computational Science – ICCS 2020*, Cham: Springer, pp. 410–424.

Gomes, C. P. (2009). Computational sustainability: Computational methods for a sustainable environment, economy, and society. *The Bridge*, **39**(4), 5–13.

González-Bailón, S. (2013). Social science in the era of big data. *Policy & Internet*, **5**(2), 147–160.

Gonzalez, M. C., Hidalgo, C. A., & Barabasi, A.-L. (2008). Understanding individual human mobility patterns. *Nature*, **453**(7196), 779–782.

Goodchild, M. F. (2007). Citizens as sensors: The world of volunteered geography. *GeoJournal*, **69**(4), 211–221.

Goodchild, M. F. (2013). The quality of big (geo)data. *Dialogues in Human Geography*, **3**(3), 280–284.

Goodchild, M. F., & Li, W. (2021). Replication across space and time must be weak in the social and environmental sciences. *Proceedings of the National Academy of Sciences*, **118**(35), e2015759118.

Gorelick, N., Hancher, M., Dixon, M. et al. (2017). Google Earth Engine: Planetary-scale geospatial analysis for everyone. *Remote Sensing of Environment*, **202**, 18–27.

Graafland, C. E., Gutiérrez, J. M., López, J. M., Pazó, D., & Rodríguez, M. A. (2020). The probabilistic backbone of data-driven complex networks: An example in climate. *Scientific Reports*, **10**(1), 11484.

Graham, M. (2018). Rethinking the Geoweb and Big Data: Future Research Directions. In J. Thatcher, A. Shears, & J. Eckert, eds., *Thinking Big Data in Geography: New Regimes, New Research*, Lincoln: University of Nebraska Press, pp. 231–236.

Graham, M., de Sabbata, S., & Zook, M. A. (2015). Towards a study of information geographies: (Im)mutable augmentations and a mapping of the geographies of information. *Geo: Geography and Environment*, **2**(1), 88–105.

Graham, M., & Shelton, T. (2013). Geography and the future of big data, big data and the future of geography. *Dialogues in Human Geography*, **3**(3), 255–261.

Grainger, A. (2009). Measuring the planet to fill terrestrial data gaps. *Proceedings of the National Academy of Sciences*, **106**(49), 20557–20558.

Granger, C. W. J. (1988). Causality, cointegration, and control. *Journal of Economic Dynamics and Control*, **12**(2), 551–559.

Graves, A., Wayne, G., Reynolds, M. et al. (2016). Hybrid computing using a neural network with dynamic external memory. *Nature*, **538**(7626), 471–476.

Gray, M. L., & Suri, S. (2017). The humans working behind the AI curtain. *Harvard Business Review*, **9**(1), 2–5.

Greenfield, A. (2010). *Everyware: The Dawning Age of Ubiquitous Computing*, Berkeley, CA: New Riders.

Gregory, M. J., Kimerling, A. J., White, D., & Sahr, K. (2008). A comparison of intercell metrics on discrete global grid systems. *Computers, Environment and Urban Systems*, **32** (3), 188–203.

Griffin, G. P., Mulhall, M., Simek, C., & Riggs, W. W. (2020). Mitigating bias in big data for transportation. *Journal of Big Data Analytics in Transportation*, **2**(1), 49–59.

Grimm, V., Berger, U., DeAngelis, D. L. et al. (2010). The ODD protocol: A review and first update. *Ecological Modelling*, **221**(23), 2760–2768.

Grimm, V., Railsback, S. F., Vincenot, C. E. et al. (2020). The ODD protocol for describing agent-based and other simulation models: A second update to improve clarity, replication, and structural realism. *Journal of Artificial Societies and Social Simulation*, **23** (2), 7.

Grolinger, K., Hayes, M., Higashino, W. A. et al. (2014). Challenges for Mapreduce in Big Data. In *2014 IEEE World Congress on Services*, Santa Clara, CA: Institute of Electrical and Electronics Engineers, pp. 182–189.

Grommé, F., Ruppert, E., & Cakici, B. (2018). Data Scientists: A New Faction of the Transnational Field of Statistics. In H. Knox & D. Nafus, eds., *Ethnography for a Data-Saturated World*, Manchester: Manchester University Press, pp. 33–61.

Guba, E. G. (1981). Criteria for assessing the trustworthiness of naturalistic inquiries. *Educational Communication and Technology*, **29**(2), 75.

Guerry, A.-M. (1833). *Essai sur la statistique morale de la France*, Paris: Arrondissements des Académies et des Cours Royales de France.

Gundersen, O. E., & Kjensmo, S. (2018). State of the art: Reproducibility in artificial intelligence. *Proceedings of the AAAI Conference on Artificial Intelligence*, **32**(1), 1644–1651.

Guo, L., & Lin, H. (2016). Critical zone research and observatories: Current status and future perspectives. *Vadose Zone Journal*, **15**(9), 1–14.

Guo, P. J. (2014). CDE: Automatically Package and Reproduce Computational Experiments. In V. Stodden, F. Leisch, & R. D. Peng, eds., *Implementing Reproducible Research*, Boca Raton, FL: CRC Press, pp. 79–112.

Gurstein, M. (2003). Effective use: A community informatics strategy beyond the digital divide. *First Monday*, **8**(12), 1107.

Habegger, B. (2010). Strategic foresight in public policy: Reviewing the experiences of the UK, Singapore, and the Netherlands. *Futures*, **42**(1), 49–58.

Hackett, E. J., Parker, J. N., Conz, D., Rhoten, D., & Parker, A. (2008). Ecology Transformed: NCEAS and Changing Patterns of Ecological Research. In *Scientific Collaboration on the Internet*, Cambridge, MA: MIT Press, pp. 277–296.

Haenlein, M., & Kaplan, A. (2019). A brief history of artificial intelligence: On the past, present, and future of artificial intelligence. *California Management Review*, **61**(4), 5–14.

Haerder, T., & Reuter, A. (1983). Principles of transaction-oriented database recovery. *ACM Computing Surveys (CSUR)*, **15**(4), 287–317.

Haklay, M. (2010). How good is volunteered geographical information? A comparative study of OpenStreetMap and Ordnance Survey datasets. *Environment and Planning B: Planning and Design*, **37**(4), 682–703.

Haklay, M. (2013). Neogeography and the delusion of democratisation. *Environment and Planning A*, **45**(1), 55–69.

Hall, J. L., Boucher, R. H., Buckland, K. N. et al. (2015). MAGI: A new high-performance airborne thermal-infrared imaging spectrometer for earth science applications. *IEEE Transactions on Geoscience and Remote Sensing*, **53**(10), 5447–5457.

Hampton, S. E., Strasser, C. A., Tewksbury, J. J. et al. (2013). Big data and the future of ecology. *Frontiers in Ecology and the Environment*, **11**(3), 156–162.

Han, J., Yang, Z., Zhang, Q. et al. (2019). A method of insulator faults detection in aerial images for high-voltage transmission lines inspection. *Applied Sciences*, **9**(10), 2009.

Hansen, M. M., Miron-Shatz, T., Lau, A. Y. S., & Paton, C. (2014). Big data in science and healthcare: A review of recent literature and perspectives. *Yearbook of Medical Informatics*, **23**(1), 21–26.

Hao, K. (2021). How Facebook and Google fund global misinformation. *MIT Technology Review*, November. Accessed August 11, 2022, www.technologyreview.com/2021/11/20/1039076/facebook-google-disinformation-clickbait/.

Hart, J. K., & Martinez, K. (2006). Environmental sensor networks: A revolution in the Earth system science? *Earth-Science Reviews*, **78**(3), 177–191.

Harvey, F. (2013). To Volunteer or to Contribute Locational Information? Towards Truth in Labeling for Crowdsourced Geographic Information. In *Crowdsourcing Geographic Knowledge*, Cham: Springer, pp. 31–42.

Hassall, C., Owen, J., & Gilbert, F. (2017). Phenological shifts in hoverflies (Diptera: Syrphidae): Linking measurement and mechanism. *Ecography*, **40**(7), 853–863.

Hastings, J., Glauer, M., Memariani, A., Neuhaus, F., & Mossakowski, T. (2021). Learning chemistry: Exploring the suitability of machine learning for the task of structure-based chemical ontology classification. *Journal of Cheminformatics*, **13**(1), 23.

Haynes, D., Ray, S., Manson, S. M., & Soni, A. (2015). High Performance Analysis of Big Spatial Data. In *Big Data: 2015 IEEE International Conference on Big Data*, Santa Clara, CA: Institute of Electrical and Electronics Engineers, pp. 1953–1957.

Henke, N., Bughin, J., Chui, M. et al. (2016). *The Age of Analytics: Competing in a Data-Driven World*. Brussels: McKinsey Global Institute.

Henrickson, L., & McKelvey, B. (2002). Foundations of "new" social science: Institutional legitimacy from philosophy, complexity science, postmodernism, and agent-based modeling. *Proceedings of the National Academy of Sciences*, **99**(90003), 7288–7295.

Hernán, M. A., Hsu, J., & Healy, B. (2019). A second chance to get causal inference right: A classification of data science tasks. *CHANCE*, **32**(1), 42–49.

Hey, A. J. G., Tansley, S., & Tolle, K. M. (2009). *The Fourth Paradigm: Data-Intensive Scientific Discovery*, Redmond, WA: Microsoft Research.

Hidalgo, C. A. (2014). Saving big data from big mouths. *Scientific American*, **311**(2), 64–67.

Higgins, S. (2008). The DCC curation lifecycle model. *Proceedings of the 8th ACM/IEEE-CS Joint Conference on Digital Libraries*, **3**(1), 453.

Hilbert, M. (2016). Big data for development: A review of promises and challenges. *Development Policy Review*, **34**(1), 135–174.

Hilbert, M., & López, P. (2011). The world's technological capacity to store, communicate, and compute information. *Science*, **332**(6025), 60–65.

Hinckley, E.-L. S., Anderson, S. P., Baron, J. S. et al. (2016). Optimizing available network resources to address questions in environmental biogeochemistry. *BioScience*, **66**(4), 317–326.

Hindman, M. (2015). Building better models: Prediction, replication, and machine learning in the social sciences. *The Annals of the American Academy of Political and Social Science*, **659**(1), 48–62.

Hoaglin, D. C., Mosteller, F., & Tukey, J. W. (1984). *Understanding Robust and Exploratory Data Analysis*, New York: John Wiley & Sons.

Hobbie, J. E., Carpenter, S. R., Grimm, N. B., Gosz, J. R., & Seastedt, T. R. (2003). The US Long Term Ecological Research Program. *BioScience*, **53**(1), 21–32.

Hochachka, W. M., Alonso, H., Gutiérrez-Expósito, C., Miller, E., & Johnston, A. (2021). Regional variation in the impacts of the COVID-19 pandemic on the quantity and quality of data collected by the project eBird. *Biological Conservation*, **254**, 108974.

Hofer, B. (2015). Uses of online geoprocessing technology in analyses and case studies: A systematic analysis of literature. *International Journal of Digital Earth*, **8**(11), 901–917.

Hogan, B. (2018). Social media giveth, social media taketh away: Facebook, friendships, and APIs. *International Journal of Communication*, **12**, 592–611.

Hollands, R. G. (2015). Critical interventions into the corporate smart city. *Cambridge Journal of Regions, Economy and Society*, **8**(1), 61–77.

Holloway, J., & Mengersen, K. (2018). Statistical machine learning methods and remote sensing for sustainable development goals: A review. *Remote Sensing*, **10**(9), 1365.

Holm, P., Goodsite, M. E., Cloetingh, S. et al. (2013). Collaboration between the natural, social and human sciences in global change research. *Environmental Science & Policy*, **28**, 25–35.

Holst, A. (2021). *Big Data*, Hamburg: Statista.

Hong, A., Kim, B., & Widener, M. (2020). Noise and the city: Leveraging crowdsourced big data to examine the spatio-temporal relationship between urban development and noise annoyance. *Environment and Planning B: Urban Analytics and City Science*, **47**(7), 1201–1218.

Houborg, R., & McCabe, M. F. (2018). A cubesat enabled spatio-temporal enhancement method (CESTEM) utilizing planet, landsat and modis data. *Remote Sensing of Environment*, **209**, 211–226.

Houser, C., Lehner, J., & Smith, A. (2022). The field geomorphologist in a time of artificial intelligence and machine learning. *Annals of the American Association of Geographers*, **112**(5), 1260–1277.

Hu, F., Yang, C., Schnase, J. L. et al. (2018). ClimateSpark: An in-memory distributed computing framework for big climate data analytics. *Computers & Geosciences*, **115**, 154–166.

Huang, Q., Cervone, G., & Zhang, G. (2017). A cloud-enabled automatic disaster analysis system of multi-sourced data streams: An example synthesizing social media, remote sensing and Wikipedia data. *Computers, Environment and Urban Systems*, **66**, 23–37.

Hulley, G. C., Duren, R. M., Hopkins, F. M. et al. (2016). High spatial resolution imaging of methane and other trace gases with the airborne Hyperspectral Thermal Emission Spectrometer (HyTES). *Atmospheric Measurement Techniques*, **9**(5), 2393–2408.

Hung, M., Lauren, E., Hon, E. S. et al. (2020). Social network analysis of COVID-19 sentiments: Application of artificial intelligence. *Journal of Medical Internet Research*, **22**(8), e22590.

Hutson, M. (2018). AI researchers allege that machine learning is alchemy. *Science*, **360** (6388), 861.

Hutton, C., Wagener, T., Freer, J. et al. (2016). Most computational hydrology is not reproducible, so is it really science? *Water Resources Research*, **52**(10), 7548–7555.

Hyvärinen, O., & Saltikoff, E. (2010). Social media as a source of meteorological observations. *Monthly Weather Review*, **138**(8), 3175–3184.

Iliadis, A., & Russo, F. (2016). Critical data studies: An introduction. *Big Data & Society*, **3** (2), https://doi.org/10.1177/2F2053951716674238.

Ilyas, M., & Mahgoub, I. (2018). *Smart Dust: Sensor Network Applications, Architecture and Design*, Boca Raton, FL: CRC press.

International Telecommunication Union (ITU). (2019). *Report on the Implementation of the Strategic Plan and the Activities of the Union for 2018–2019*, Geneva: ITU.

International Telecommunication Union (ITU). (2021). *Measuring Digital Development: Facts and figures 2021*, Geneva: ITU.

Ioannidis, J. P. A. (2008). Measuring co-authorship and networking-adjusted scientific impact. *PLOS ONE*, 3(7), e2778.

Ioannidis, J. P. A. (2013). Informed consent, big data, and the oxymoron of research that is not research. *The American Journal of Bioethics*, **13**(4), 40–42.

Isdahl, R., & Gundersen, O. E. (2019). Out-of-the-Box Reproducibility: A Survey of Machine Learning Platforms. In *2019 15th International Conference on eScience (eScience)*, San Diego, CA: Institute of Electrical and Electronics Engineers, pp. 86–95.

Ishwarappa, & Anuradha, J. (2015). A brief introduction on big data 5Vs characteristics and hadoop technology. *Procedia Computer Science*, **48**, 319–324.

Jacobs, A. (2009). The pathologies of big data. *Communications of the ACM*, **52**(8), 36–44.

Jacobs, N., Burgin, W., Fridrich, N. et al. (2009). The Global Network of Outdoor Webcams: Properties and Applications. In *Proceedings of the 17th ACM SIGSPATIAL International Conference on Advances in Geographic Information Systems*, New York: Association for Computing Machinery, pp. 111–120.

Jagadish, H. V. (2015). Big data and science: Myths and reality. *Big Data Research*, **2**(2), 49–52.

Jain, P., Gyanchandani, M., & Khare, N. (2016). Big data privacy: A technological perspective and review. *Journal of Big Data*, **3**(1), 25.

Janowicz, K., Gao, S., McKenzie, G., Hu, Y., & Bhaduri, B. (2020). GeoAI: Spatially explicit artificial intelligence techniques for geographic knowledge discovery and beyond. *International Journal of Geographical Information Science*, **34**(4), 625–636.

Janowicz, K., Hitzler, P., Li, W. et al. (2022). Know, know where, KnowWhereGraph: A densely connected, cross-domain knowledge graph and geo-enrichment service stack for applications in environmental intelligence. *AI*, **43**(1), 30–39.

Janssen, M. A. (2017). The practice of archiving model code of agent-based models. *Journal of Artificial Societies and Social Simulation*, **20**(1), 2.

Janssen, M., Estevez, E., & Janowski, T. (2014). Interoperability in big, open, and linked data: Organizational maturity, capabilities, and data portfolios. *IEEE Annals of the History of Computing*, **47**(10), 44–49.

Japkowicz, N., & Stephen, S. (2002). The class imbalance problem: A systematic study. *Intelligent Data Analysis*, **6**(5), 429–449.

Jean, N., Burke, M., Xie, M. et al. (2016). Combining satellite imagery and machine learning to predict poverty. *Science*, **353**(6301), 790–794.

Jefferson, B. (2020). *Digitize and Punish: Racial Criminalization in the Digital Age*, Minneapolis: University of Minnesota Press.

Jeffery, K., Pursula, A., & Zhao, Z. (2020). ICT Infrastructures for Environmental and Earth Sciences. In Z. Zhao & M. Hellström, eds., *Towards Interoperable Research Infrastructures for Environmental and Earth Sciences*, New York: Springer, pp. 17–29.

Jensen, J. R. (1986). *Introductory Digital Image Processing: A Remote Sensing Perspective*, Columbus: University of South Carolina Press.

Jensen, R. (2007). The digital provide: Information (technology), market performance, and welfare in the South Indian fisheries sector. *The Quarterly Journal of Economics*, **122**(3), 879–924.

Johnson, K. (2022). The Census Is Broken. Can AI Fix It? *Wired*, April 8. Accessed July 23, 2022, www.wired.com/story/us-census-undercount-ai-satellites.

Jones, P., Drury, R., & McBeath, J. (2011). Using GPS-enabled mobile computing to augment qualitative interviewing: Two case studies. *Field Methods*, **23**(2), 173–187.

Jongman, B., Wagemaker, J., Romero, R. B., & De Perez, C. E. (2015). Early flood detection for rapid humanitarian response: Harnessing near real-time satellite and Twitter signals. *ISPRS International Journal of Geo-Information*, **4**(4), 2246–2266.

Joppa, L. N. (2017). The case for technology investments in the environment. *Nature*, **552**, 325–329.

Jurado Lozano, P. J., & Regan, A. (2018). Land Surface Satellite Remote Sensing Gap Analysis. In K. Themistocleous, D. G. Hadjimitsis, S. Michaelides, V. Ambrosia, & G. Papadavid, eds., *Sixth International Conference on Remote Sensing and Geoinformation of the Environment (RSCy2018)*, Paphos: SPIE, p. 10773.

Kalil, T. (2012). Big Data Is a Big Deal. Accessed September 1, 2019, https://obamawhite house.archives.gov/blog/2012/03/29/big-data-big-deal.

Kamilaris, A., Kartakoullis, A., & Prenafeta-Boldú, F. X. (2017). A review on the practice of big data analysis in agriculture. *Computers and Electronics in Agriculture*, **143**, 23–37.

Kandt, J., & Batty, M. (2021). Smart cities, big data and urban policy: Towards urban analytics for the long run. *Cities*, **109**, 102992.

Kaplan, F. (2016). *Dark Territory: The Secret History of Cyber War*, New York: Simon and Schuster.

Karasti, H., & Baker, K. S. (2008). Digital data practices and the long term ecological research program growing global. *International Journal of Digital Curation*, **3**(2), 42–58.

Karpatne, A., Atluri, G., Faghmous, J. H. et al. (2017). Theory-guided data science: A new paradigm for scientific discovery from data. *IEEE Transactions on Knowledge and Data Engineering*, **29**(10), 2318–2331.

Katal, A., Wazid, M., & Goudar, R. H. (2013). Big data: Issues, challenges, tools and Good practices. In *2013 Sixth International Conference on Contemporary Computing (IC3)*, Santa Clara, CA: Institute of Electrical and Electronics Engineers, pp. 404–409.

Katz, L. (2019). *Evaluation of the Moore–Sloan Data Science Environments*, New York: Alfred P. Sloan Foundation.

Kawale, J., Chatterjee, S., Ormsby, D. et al. (2012). Testing the Significance of Spatio-Temporal Teleconnection Patterns. In *Proceedings of the 18th ACM SIGKDD International Conference on Knowledge Discovery and Data Mining – KDD '12*, Beijing: Association for Computing Machinery, pp. 642–650.

Kedron, P., Frazier, A. E., Trgovac, A. B., Nelson, T., & Fotheringham, A. S. (2021). Reproducibility and replicability in geographical analysis. *Geographical Analysis*, **53** (1), 135–147.

Keeler, B. L., Wood, S. A., Polasky, S. et al. (2015). Recreational demand for clean water: Evidence from geotagged photographs by visitors to lakes. *Frontiers in Ecology and the Environment*, **13**(2), 76–81.

Kelley, M. (2018). Framing Digital Exclusion in Technologically Mediated Urban Spaces. In J. Thatcher, A. Shears, & J. Eckert, eds., *Thinking Big Data in Geography: New Regimes, New Research*, Lincoln: University of Nebraska Press, pp. 178–193.

Kelling, S., Hochachka, W. M., Fink, D. et al. (2009). Data-intensive science: A new paradigm for biodiversity studies. *BioScience*, **59**(7), 613–620.

Khoury, M. J., & Ioannidis, J. P. A. (2014). Big data meets public health. *Science*, **346** (6213), 1054–1055.

Kientz, J. A. (2019). In praise of small data: When you might consider N-of-1 studies. *GetMobile: Mobile Computing and Communications*, **22**(4), 5–8.

Kim, G.-H., Trimi, S., & Chung, J.-H. (2014). Big-data applications in the government sector. *Communications of the ACM*, **57**(3), 78–85.

King, E., & Martin, N. (2015). *Turning Data into Action*, Washington, DC: US Agency for International Development.

King, G. (2011). Ensuring the data-rich future of the social sciences. *Science*, **331**(6018), 719–721.

Kirkpatrick, R. (2012). Big data for development. *Big Data*, **1**(1), 3–4.

Kitchin, R. (2014a). *The Data Revolution: Big Data, Open Data, Data Infrastructures & Their Consequences*, Newbury Park, CA: Sage Publications.

Kitchin, R. (2014b). The real-time city? Big data and smart urbanism. *GeoJournal*, **79** (1), 1–14.

Kitchin, R., & Dodge, M. (2011). *Code/Space: Software and Everyday Life*, Cambridge, MA: MIT Press.

Kitchin, R., & Lauriault, T. P. (2015). Small data in the era of big data. *GeoJournal*, **80**(4), 463–475.

Kitchin, R., & McArdle, G. (2016). What makes big data, big data? Exploring the ontological characteristics of 26 datasets. *Big Data & Society*, **3**(1), https://doi.org/10.1177/2053951716631130.

Kitzes, J., Turek, D., & Deniz, F. (2017). *The Practice of Reproducible Research: Case Studies and Lessons from the Data-Intensive Sciences*, Oakland: University of California Press.

Klemens, B. (2021). Keeping science reproducible in a world of custom code and data. *Ars Technica*. Accessed https://arstechnica.com/science/2021/11/keeping-science-reproducible-in-a-world-of-custom-code-and-data/. Code and data: https://zenodo.org/badge/latestdoi/412662986.

Klocke, D., Pincus, R., & Quaas, J. (2011). On constraining estimates of climate sensitivity with present-day observations through model weighting. *Journal of Climate*, **24**(23), 6092–6099.

Knight, W. (2017). The dark secret at the heart of AI. *Technology Review*, **120**(3), 54–61.

Knox, H., & Nafus, D. (2018). Introduction: Ethnography for a Data-Saturated World. In H. Knox & D. Nafus, eds., *Ethnography for a Data-Saturated World*, Manchester: Manchester University Press, pp. 1–30.

Kohler, R. E. (2002). Place and practice in field biology. *History of Science*, **40**(2), 189–210.

König, C., Weigelt, P., Schrader, J. et al. (2019). Biodiversity data integration: The significance of data resolution and domain. *PLOS Biology*, **17**(3), e3000183.

Kortuem, G., Kawsar, F., Sundramoorthy, V., & Fitton, D. (2010). Smart objects as building blocks for the internet of things. *IEEE Internet Computing*, **14**(1), 44–51.

Kotsev, A., Pantisano, F., Schade, S., & Jirka, S. (2015). Architecture of a service-enabled sensing platform for the environment. *Sensors*, **15**(2), 4470–4495.

Kreinovich, V., & Ouncharoen, R. (2015). Fuzzy (and interval) techniques in the age of big data: An overview with applications to environmental science, geosciences, engineering, and medicine. *International Journal of Uncertainty, Fuzziness and Knowledge-Based Systems*, **23**(suppl. 1), 75–89.

Krishnan, R., Samaranayake, V. A., & Jagannathan, S. (2019). A hierarchical dimension reduction approach for big data with application to fault diagnostics. *Big Data Research*, **18**, 100121.

Kroll, J. A., Barocas, S., Felten, E. W. et al. (2016). Accountable algorithms. *University of Pennsylvania Law Review*, **165**, 633.

Krzyzanowski, B., & Manson, S. M. (2022). Regionalization with self-organizing maps for sharing higher resolution protected health information. *Annals of the Association of American Geographers*, **112**, https://doi.org/10.1080/24694452.2021.2020617.

Kshetri, N. (2014). The emerging role of big data in key development issues: Opportunities, challenges, and concerns. *Big Data & Society*, **1**(2), https://doi.org/10.1177/2053951714564227.

Kugler, T. A., & Fitch, C. A. (2018). Interoperable and accessible census and survey data from IPUMS. *Scientific Data*, **5**(1), 180007.

Kugler, T. A., Grace, K., Wrathall, D. J. et al. (2019). People and pixels 20 years later: The current data landscape and research trends blending population and environmental data. *Population and Environment*, **41**, 209–234.

Kugler, T. A., Manson, S. M., & Donato, J. R. (2017). Spatiotemporal aggregation for temporally extensive international microdata. *Computers, Environment and Urban Systems*, **63**, 26–37.

Kugler, T. A., van Riper, D. C. D. C., Manson, S. M. et al. (2015). *Terra populus*: Workflows for integrating and harmonizing geospatial population and environmental data. *Journal of Map & Geography Libraries*, **11**(2), 180–206.

Kulmala, M. (2018). Build a global Earth observatory. *Nature*, **553**, 21–23.

L'Heureux, A., Grolinger, K., Elyamany, H. F., & Capretz, M. A. M. (2017). Machine learning with big data: Challenges and approaches. *IEEE Access*, **5**, 7776–7797.

La Sorte, F. A., Lepczyk, C. A., Burnett, J. L. et al. (2018). Opportunities and challenges for big data ornithology. *The Condor*, **120**(2), 414–426.

Ladino, J. K. (2018). What is missing? An affective digital environmental humanities. *Resilience: A Journal of the Environmental Humanities*, **5**(2), 189–211.

Lake, B. M., Ullman, T. D., Tenenbaum, J. B., & Gershman, S. J. (2017). Building machines that learn and think like people. *Behavioral and Brain Sciences*, **40**, E253.

Landefeld, S. (2014). Uses of Big Data for Official Statistics: Privacy, Incentives, Statistical Challenges, and Other Issues. In *International Conference on Big Data for Official Statistics*, Beijing: United Nations Global Working Group (GWG) on Big Data for Official Statistics, pp. 28–30.

Lane, N. D., Miluzzo, E., Lu, H. et al. (2010). A survey of mobile phone sensing. *IEEE Communications*, **48**(9), 140–150.

Laney, D. (2001). 3D data management: Controlling data volume, velocity and variety. *META Group Research Note*, **6**(70), 949.

Lappalainen, H. K., Petäjä, T., Kujansuu, J. et al. (2014). Pan Eurasian Experiment (PEEX): A research initiative meeting the grand challenges of the changing environment of the northern pan-Eurasian arctic-boreal areas. *Geography, Environment, Sustainability*, **7**(2), 13–48.

LaRue, M. A., Stapleton, S., Porter, C. et al. (2015). Testing methods for using high-resolution satellite imagery to monitor polar bear abundance and distribution. *Wildlife Society Bulletin*, **39**(4), 772–779.

Laso Bayas, J. C., Lesiv, M., Waldner, F. et al. (2017). A global reference database of crowdsourced cropland data collected using the Geo-Wiki platform. *Scientific Data*, **4**(1), 170136.

Latour, B. (1986). Visualisation and cognition: Drawing things together. *Knowledge and Society Studies in the Sociology of Culture Past and Present*, **6**(1), 1–40.

Lauriault, T. P., Craig, B. L., Taylor, D. R. F., & Pulsifer, P. L. (2007). Today's data are part of tomorrow's research: Archival issues in the sciences. *Archivaria*, **64**, 123–179.

Lautenbacher, C. C. (2006). The Global Earth Observation System of Systems: Science serving society. *Space Policy*, **22**(1), 8–11.

Lauvaux, T., Giron, C., Mazzolini, M. et al. (2022). Global assessment of oil and gas methane ultra-emitters. *Science*, **375**(6580), 557–561.

Lazaroiu, G. C., & Roscia, M. (2012). Definition methodology for the smart cities model. *Energy*, **47**(1), 326–332.

Lazer, D., Kennedy, R., King, G., & Vespignani, A. (2014). The parable of Google Flu: Traps in big data analysis. *Science*, **343**(6176), 1203–1205.

Lazer, D., Pentland, A., Adamic, L. et al. (2009). Computational social science. *Science*, **323**(5915), 721–723.

Le Boursicaud, R., Pénard, L., Hauet, A., Thollet, F., & Le Coz, J. (2016). Gauging extreme floods on YouTube: Application of LSPIV to home movies for the post-event determination of stream discharges. *Hydrological Processes*, **30**(1), 90–105.

Leavitt, N. (2010). Will NoSQL databases live up to their promise? *Computer*, **43**(2), 12–14.

Lecocq, T., Hicks, S. P., Van Noten, K. et al. (2020). Global quieting of high-frequency seismic noise due to COVID-19 pandemic lockdown measures. *Science*, **369**(6509), 1338–1343.

Lee, D. B. (1973). A requiem for large-scale models. *Journal of the American Institute of Planners*, **39**, 163–178.

Lee, E. M. J., & O'Malley, K. G. (2020). Big fishery, big data, and little crabs: Using genomic methods to examine the seasonal recruitment patterns of early life stage dungeness crab (*Cancer magister*) in the California current ecosystem. *Frontiers in Marine Science*, **6**, 836.

Lee, H., Seo, B., Koellner, T., & Lautenbach, S. (2019). Mapping cultural ecosystem services 2.0: Potential and shortcomings from unlabeled crowd sourced images. *Ecological Indicators*, **96**, 505–515.

Lehning, M., Dawes, N., Bavay, M. et al. (2009). Instrumenting the Earth: Next Generation Sensor Networks in Environmental Science. In *The Fourth Paradigm: Data-Intensive Scientific Discovery*, Redmond, WA: Microsoft Corporation, pp. 45–51.

Leonelli, S. (2014). What difference does quantity make? On the epistemology of big data in biology. *Big Data & Society*, **1**(1), https://doi.org/10.1177/2053951714534395.

Leonelli, S. (2019). Data governance is key to interpretation: Reconceptualizing data in data science. *Harvard Data Science Review*, **1**(1), https://doi.org/10.1162/99608f92 .17405bb6.

Lerman, J. (2013). Big data and its exclusions. *Stanford Law Review Online*, **66**, 55–63.

Leszczynski, A. (2015). Spatial big data and anxieties of control. *Environment and Planning D: Society and Space*, **33**(6), 965–984.

Letouzé, E. (2012). *Big Data for Development: Challenges & Opportunities*, New York: Global Pulse.

Letouzé, E., & Jütting, J. (2015). *Official Statistics, Big Data, and Human Development*, New York: Data-Pop Alliance.

Levin, N., Ali, S., Crandall, D., & Kark, S. (2019). World heritage in danger: Big data and remote sensing can help protect sites in conflict zones. *Global Environmental Change*, **55**, 97–104.

Levin, N., Kark, S., & Crandall, D. (2015). Where have all the people gone? Enhancing global conservation using night lights and social media. *Ecological Applications*, **25**(8), 2153–2167.

Levitt, J. (2019). Citizenship and the census. *Columbia Law Review*, **119**(5), 1355–1398.

Leyk, S., Gaughan, A. E., Adamo, S. B. et al. (2019). The spatial allocation of population: A review of large-scale gridded population data products and their fitness for use. *Earth System Science Data*, **11**(3), 1385–1409.

Li, L., Goodchild, M. F., & Xu, B. (2013). Spatial, temporal, and socioeconomic patterns in the use of Twitter and Flickr. *Cartography and Geographic Information Science*, **40**(2), 61–77.

Li, W. (2020a). GeoAI: Where machine learning and big data converge in GIScience. *Journal of Spatial Information Science*, **20**, 71–77.

Li, Y., Eldawy, A., Xue, J. et al. (2019). Scalable computational geometry in MapReduce. *The VLDB Journal*, **28**(4), 523–548.

Li, Z. (2020b). Geospatial Big Data Handling with High Performance Computing: Current Approaches and Future Directions. In W. Tang & S. Wang, eds., *High Performance Computing for Geospatial Applications*, Cham: Springer, pp. 53–76.

Li, Z., Hu, F., Schnase, J. L. et al. (2017). A spatiotemporal indexing approach for efficient processing of big array-based climate data with MapReduce. *International Journal of Geographical Information Science*, **31**(1), 17–35.

Liakos, K., Busato, P., Moshou, D., Pearson, S., & Bochtis, D. (2018). Machine learning in agriculture: A review. *Sensors*, **18**(8), 2674.

Liang, F., Das, V., Kostyuk, N., & Hussain, M. M. (2018). Constructing a data-driven society: China's social credit system as a state surveillance infrastructure. *Policy & Internet*, **10**(4), 415–453.

Liang, P., & Viegas, E. (2015). CodaLab Worksheets for Reproducible, Executable Papers. In *Twenty-Ninth Conference on Neural Information Processing Systems*, Montreal: Neural Information Processing Systems.

Lin, J. (2015). On building better mousetraps and understanding the human condition: Reflections on big data in the social sciences. *The Annals of the American Academy of Political and Social Science*, **659**(1), 33–47.

Ling, F., & Foody, G. M. (2019). Super-resolution land cover mapping by deep learning. *Remote Sensing Letters*, **10**(6), 598–606.

Liu, J. C.-E., & Zhao, B. (2017). Who speaks for climate change in China? Evidence from Weibo. *Climatic Change*, **140**(3), 413–422.

Liu, Z., Guo, H., & Wang, C. (2016). Considerations on geospatial big data. *IOP Conference Series: Earth and Environmental Science*, **46**(1), 012058.

Lo, C. P., Quattrochi, D. A., & Luvall, J. C. (1997). Application of high-resolution thermal infrared remote sensing and GIS to assess the urban heat island effect. *International Journal of Remote Sensing*, **18**(2), 287–304.

Loglisci, C., & Malerba, D. (2017). Leveraging temporal autocorrelation of historical data for improving accuracy in network regression. *Statistical Analysis and Data Mining: The ASA Data Science Journal*, **10**(1), 40–53.

Lokers, R., Knapen, R., Janssen, S., van Randen, Y., & Jansen, J. (2016). Analysis of big data technologies for use in agro-environmental science. *Environmental Modelling & Software*, **84**, 494–504.

Lomazzi, M., Borisch, B., & Laaser, U. (2014). The Millennium Development Goals: Experiences, achievements and what's next. *Global Health Action*, **7**(1), 23695.

Lopes, C., & Bailur, S. (2018). *Gender Equality and Big Data: Making Gender Data Visible*, New York: United Nations Women Innovation Facility.

Lövbrand, E., Stripple, J., & Wiman, B. (2009). Earth system governmentality: Reflections on science in the Anthropocene. *Global Environmental Change*, **19**(1), 7–13.

Loveland, T. R., Sohl, T. L., Stehman, S. V et al. (2002). A strategy for estimating the rates of recent United States land-cover changes. *Photogrammetric Engineering and Remote Sensing*, **68**(10), 1091–1099.

Lu, X., Wrathall, D. J., Sundsøy, P. R. et al. (2016). Unveiling hidden migration and mobility patterns in climate stressed regions: A longitudinal study of six million anonymous mobile phone users in Bangladesh. *Global Environmental Change*, **38**, 1–7.

Ludäscher, B., Altintas, I., Berkley, C. et al. (2006). Scientific workflow management and the Kepler system. *Concurrency and Computation: Practice and Experience*, **18**(10), 1039–1065.

Lwasa, S. (2015). A systematic review of research on climate change adaptation policy and practice in Africa and South Asia deltas. *Regional Environmental Change*, **15**(5), 815–824.

Lyon, D. (2001). *Surveillance Society: Monitoring Everyday Life*, Philadelphia: Open University Press.

Maalsen, S., & Perng, S.-Y. (2016). Encountering the City at Hacking Events. In R. Kitchin & S.-Y. Perng, eds., *Code and the City*, Abingdon: Routledge, pp. 190–199.

Machlup, F. (1962). *The Production and Distribution of Knowledge in the United States*, Princeton, NJ: Princeton University Press.

MacKerron, G., & Mourato, S. (2013). Happiness is greater in natural environments. *Global Environmental Change*, **23**(5), 992–1000.

Maeda, E. E., & Torres, J. A. (2012). Open environmental data in developing countries: Who benefits? *Ambio*, **41**(4), 410–412.

Magnuson, J. J., & Bowser, C. J. (1990). A network for long-term ecological research in the United States. *Freshwater Biology*, **23**(1), 137–143.

Mahrenbach, L. C., Mayer, K., & Pfeffer, J. (2018). Policy visions of big data: Views from the Global South. *Third World Quarterly*, **39**(10), 1861–1882.

Mair, S., Jones, A., Ward, J. et al. (2018). A Critical Review of the Role of Indicators in Implementing the Sustainable Development Goals. In W. Leal Filho, ed., *Handbook of Sustainability Science and Research*, Cham: Springer, pp. 41–56.

Mankiewicz, R. (2000). *The Story of Mathematics*, Princeton, NJ: Princeton University Press.

Mann, S., Nolan, J., & Wellman, B. (2003). Sousveillance: Inventing and using wearable computing devices for data collection in surveillance environments. *Surveillance & Society*, **1**(3), 331–355.

Mannarswamy, S., & Roy, S. (2018). Evolving AI from Research to Real Life: Some Challenges and Suggestions. In *Proceedings of the Twenty-Seventh International Joint Conference on Artificial Intelligence*, Stockholm: International Joint Conferences on Artificial Intelligence Organization, pp. 5172–5179.

Manovich, L. (2011). Trending: The Promises and the Challenges of Big Social Data. In M. K. Gold, ed., *Debates in the Digital Humanities*, vol. 2, Minneapolis: University of Minnesota Press, pp. 460–475.

Manson, S., An, L., Clarke, K. et al. (2020a). Methodological issues of spatial agent-based models. *Journal of Artificial Societies and Social Simulation*, **23**(1), 3.

Manson, S. M. (2007). Challenges in evaluating models of geographic complexity. *Environment and Planning B*, **34**(2), 245–260.

Manson, S. M. (2008). Does scale exist? An epistemological scale continuum for complex human–environment systems. *Geoforum*, **39**(2), 776–788.

Manson, S. M. (2015). Digital Computer. In M. Monmonier, ed., *Cartography in the Twentieth Century*, vol. 6, Chicago: University of Chicago Press, pp. 269–270.

Manson, S. M., & O'Sullivan, D. (2006). Complexity theory in the study of space and place. *Environment and Planning A*, **38**(4), 677–692.

Manson, S. M., Schroeder, J., van Riper, D., & Ruggles, S. (2020b). *IPUMS NHGIS: VERSION 15.0*, Minneapolis, MN: Institute for Social Research and Data Innovation, http://doi.org/10.18128/D050.V15.0.

Manyika, J., Chui, M., Brown, B. et al. (2011). *Big Data: The Next Frontier for Innovation, Competition, and Productivity*, San Francisco: McKinsey Global Institute.

Marcus, G. (2018). Deep learning: A critical appraisal. *ArXiv Preprint*, ArXiv:1801.00631.

Mardani, A., Liao, H., Nilashi, M., Alrasheedi, M., & Cavallaro, F. (2020). A multi-stage method to predict carbon dioxide emissions using dimensionality reduction, clustering, and machine learning techniques. *Journal of Cleaner Production*, **275**, 122942.

Markham, A. N. (2018). Afterword: Ethics as impact: Moving from error-avoidance and concept-driven models to a future-oriented approach. *Social Media + Society*, **4**(3), https://doi.org/10.1177/2056305118784504.

Marsh, G. P. (1864). *Man and Nature or, Physical Geography as Modified by Human Action*, New York: Charles Scribner.

Marton, A., Avital, M., & Jensen, T. B. (2013). Reframing Open Big Data. In *ECIS 2013 – Proceedings of the 21st European Conference on Information Systems (2013)*, Utrecht: Association for Information Systems, p. 146.

Marz, N., & Warren, J. (2015). *Big Data: Principles and Best Practices of Scalable Real-Time Data Systems*, New York: Manning Publications.

Masmoudi, M., Karray, M. H., Ben Abdallah Ben Lamine, S., Zghal, H. B., & Archimede, B. (2020). MEMOn: Modular Environmental Monitoring Ontology to link heterogeneous Earth observed data. *Environmental Modelling & Software*, **124**, 104581.

Mateus, C., Potito, A., & Curley, M. (2021). Engaging secondary school students in climate data rescue through service-learning partnerships. *Weather*, **76**(4), 113–118.

Matsuoka, S., Sato, H., Tatebe, O. et al. (2014). Extreme big data (EBD): Next generation big data infrastructure technologies towards yottabyte/year. *Supercomputing Frontiers and Innovations*, **1**(2), 89–107.

Mattern, S. (2013). Methodolatry and the art of measure. Places Journal, November. Accessed September 4, 2022, https://doi.org/10.22269/131105.

Matthew, H. (2018). Artificial intelligence faces reproducibility crisis. *Science*, **359**(6377), 725–726.

Maxwell, A. E., Warner, T. A., & Fang, F. (2018). Implementation of machine-learning classification in remote sensing: an applied review. *International Journal of Remote Sensing*, **39**(9), 2784–2817.

Maybury, M. T. (2012). *Multimedia Information Extraction: Advances in Video, Audio, and Imagery Analysis for Search, Data Mining, Surveillance and Authoring*, Hoboken, NJ: John Wiley & Sons.

Mayer-Schönberger, V., & Cukier, K. (2013). *Big Data: A Revolution That Will Transform How We Live, Work, and Think*, New York: Houghton Mifflin Harcourt.

Mayernik, M. S., Wallis, J. C., & Borgman, C. L. (2013). Unearthing the infrastructure: Humans and sensors in field-based scientific research. *Computer Supported Cooperative Work (CSCW)*, **22**(1), 65–101.

Mayson, S. G. (2018). Bias in, bias out. *Yale Law Journal*, **128**, 2218–2300.

McBride, B. (2004). The Resource Description Framework (RDF) and Its Vocabulary Description Language RDFS. In *Handbook on Ontologies*, Heidelberg: Springer, pp. 51–65.

McCarthy, M. T. (2016). The big data divide and its consequences. *Sociology Compass*, **10** (12), 1131–1140.

McCulloch, W. S., & Pitts, W. (1943). A logical calculus of the ideas imminent in nervous activity. *Bulletin of Mathematical Biophysics*, **5**, 115–137.

McDermott, P. (2010). Building open government. *Government Information Quarterly*, **27** (4), 401–413.

McDuie-Ra, D., & Gulson, K. (2020). The backroads of AI: The uneven geographies of artificial intelligence and development. *Area*, **52**(3), 626–633.

McGeer, R., Berman, M., Elliott, C., & Ricci, R. (2016). *The GENI Book*, Cham: Springer.

McGovern, A., Elmore, K. L., Gagne, D. J. et al. (2017). Using artificial intelligence to improve real-time decision-making for high-impact weather. *Bulletin of the American Meteorological Society*, **98**(10), 2073–2090.

McGregor, C., & Bonnis, B. (2017). New approaches for integration: Integration of haptic garments, big data analytics, and serious games for extreme environments. *IEEE Consumer Electronics Magazine*, **6**(4), 92–96.

McNulty, S. A., White, D., Hufty, M., & Foster, P. (2017). The Organization of Biological Field Stations at fifty. *Bulletin of the Ecological Society of America*, **98**(4), 359–373.

Meier, P. (2015). *Digital Humanitarians: How Big Data Is Changing the Face of Humanitarian Response*, Abingdon: Routledge.

Meinshausen, N., Hauser, A., Mooij, J. M. et al. (2016). Methods for causal inference from gene perturbation experiments and validation. *Proceedings of the National Academy of Sciences*, **113**(27), 7361–7368.

Meng, X.-L. (2018). Statistical paradises and paradoxes in big data (I): Law of large populations, big data paradox, and the 2016 US presidential election. *The Annals of Applied Statistics*, **12**(2), 685–726.

Meyer, H., & Pebesma, E. (2022). Machine learning-based global maps of ecological variables and the challenge of assessing them. *Nature Communications*, **13**, 2208.

Miceli, M., Posada, J., & Yang, T. (2022). Studying up machine learning data: Why talk about bias when we mean power? *Proceedings of the ACM on Human-Computer Interaction*, **6**, 34.

Michener, W. K., Allard, S., Budden, A. et al. (2012). Participatory design of DataONE: Enabling cyberinfrastructure for the biological and environmental sciences. *Ecological Informatics*, **11**, 5–15.

Milakis, D., van Arem, B., & Van Wee, B. (2017). Policy and society related implications of automated driving: A review of literature and directions for future research. *Journal of Intelligent Transportation Systems*, **21**(4), 324–348.

Miller, H. G., & Mork, P. (2013). From data to decisions: A value chain for big data. *IT Professional*, **15**(1), 57–59.

Miller, H. J., & Goodchild, M. F. (2015). Data-driven geography. *GeoJournal*, **80**(4), 449–461.

Millett, L. I., & Estrin, D. L. (2012). *Computing Research for Sustainability*, Washington, DC: National Academies Press.

Mishelevich, D. J. (1986). Artificial intelligence in medicine: The commercial realities. *Software in Healthcare*, **4**(3), 28–30.

Mohri, M., Rostamizadeh, A., & Talwalkar, A. (2018). *Foundations of Machine Learning*, Cambridge, MA: MIT press.

Mojaddadi, H., Pradhan, B., Nampak, H., Ahmad, N., & Ghazali, A. H. bin. (2017). Ensemble machine-learning-based geospatial approach for flood risk assessment using multi-sensor remote-sensing data and GIS. *Geomatics, Natural Hazards and Risk*, **8**(2), 1080–1102.

Moltmann, T., Turton, J., Zhang, H.-M. et al. (2019). A Global Ocean Observing System (GOOS), delivered through enhanced collaboration across regions, communities, and new technologies. *Frontiers in Marine Science*, **6**, 291.

Monajemi, H., Murri, R., Jonas, E. et al. (2019). Ambitious data science can be painless. *Harvard Data Science Review*, **1**(1), https://doi.org/10.1162/99608f92.02ffc552.

Monroe, B. L., Pan, J., Roberts, M. E., Sen, M., & Sinclair, B. (2015). No! Formal theory, causal inference, and big data are not contradictory trends in political science. *PS: Political Science & Politics*, **48**(1), 71–74.

Mooney, H. A., Duraiappah, A., & Larigauderie, A. (2013). Evolution of natural and social science interactions in global change research programs. *Proceedings of the National Academy of Sciences*, **110**(suppl. 1), 3665–3672.

Mora, L., Deakin, M., Zhang, X. et al. (2020). Assembling sustainable smart city transitions: An interdisciplinary theoretical perspective. *Journal of Urban Technology*, **28**(1–2), 1–27.

Moran, E. F., & Lopez, M. C. (2016). Future directions in human–environment research. *Environmental Research*, **144**(B), 1–7.

Moretti, F. (2000). Conjectures on world literature. *New Left Review*, **1**(1), 54–68.

Moretti, F. (2005). *Graphs, Maps, Trees: Abstract Models for a Literary History*, New York: Verso.

Morgan, S. L., & Winship, C. (2014). *Counterfactuals and Causal Inference Methods and Principles for Social Research*, Cambridge: Cambridge University Press.

Morozov, E. (2013). *To Save Everything, Click Here: The Folly of Technological Solutionism*, New York: Public Affairs.

Morrow, N., Mock, N., Papendieck, A., & Kocmich, N. (2011). *Independent Evaluation of the Ushahidi Haiti Project*, New Orleans, FL: Development Information Systems International.

Morsy, M. M., Goodall, J. L., Castronova, A. M. et al. (2017). Design of a metadata framework for environmental models with an example hydrologic application in HydroShare. *Environmental Modelling and Software*, **93**, 13–28.

Moses, M. E. (2013). Information technology: Slouching towards utopia. *Nature*, **502** (7471), 299–300.

Mueller, J. T., & Santos-Lozada, A. R. (2022). The 2020 US Census differential privacy method introduces disproportionate discrepancies for rural and non-white populations. *Population Research and Policy Review*, **41**, 1417–1430.

Müller, B., Bohn, F., Dreßler, G. et al. (2013). Describing human decisions in agent-based models–ODD+ D, an extension of the ODD protocol. *Environmental Modelling & Software*, **48**, 37–48.

Muller, C. L., Chapman, L., Johnston, S. et al. (2015). Crowdsourcing for climate and atmospheric sciences: current status and future potential. *International Journal of Climatology*, **35**(11), 3185–3203.

Munafò, M. R., Nosek, B. A., Bishop, D. V. M. et al. (2017). A manifesto for reproducible science. *Nature Human Behaviour*, **1**, 21.

Muñoz, L., Hausner, V. H., Runge, C., Brown, G., & Daigle, R. (2020). Using crowdsourced spatial data from Flickr vs. PPGIS for understanding nature's contribution to people in Southern Norway. *People and Nature*, **2**(2), 437–449.

Musen, M. A., Bean, C. A., Cheung, K.-H. et al. (2015). The center for expanded data annotation and retrieval. *Journal of the American Medical Informatics Association: JAMIA*, **22**(6), 1148–1152.

Nadon, G., Feilberg, M., Johansen, M., & Shklovski, I. (2018). In the User We Trust: Unrealistic Expectations of Facebook's Privacy Mechanisms. In *Proceedings of the 9th International Conference on Social Media and Society*, New York: Association for Computing Machinery, pp. 138–149.

Naidoo, R., & Fisher, B. (2020). Reset Sustainable Development Goals for a pandemic world. *Nature*, **583**(7815), 198–201.

Nakagawa, S., & Parker, T. H. (2015). Replicating research in ecology and evolution: Feasibility, incentives, and the cost-benefit conundrum. *BMC Biology*, **13**(1), 88.

NASA. (2020). *NASA's Management of Distributed Active Archive Centers*, Washington, DC: NASA.

NASEM. (2018). *Data Science for Undergraduates: Opportunities and Options*, Washington, DC: National Academies Press.

NASEM. (2019). *Reproducibility and Replicability in Science*, Washington, DC: National Academies Press. www.nap.edu/catalog/25303/reproducibility-and-replicability-in-science.

National Research Council (NRC). (2015). *Training Students to Extract Value from Big Data: Summary of a Workshop*, Washington, DC: National Academies Press.

Nativi, S., Mazzetti, P., Santoro, M. et al. (2015). Big data challenges in building the global earth observation system of systems. *Environmental Modelling & Software*, **68**, 1–26.

Naur, P. (1974). *Concise Survey of Computer Methods*, Lund: Studentlitteratur.

Ndung'u, N. (2018). The M-Pesa Technological Revolution for Financial Services in Kenya: A Platform for Financial Inclusion. In D. Lee & R. H. Deng, eds., *Handbook of Blockchain, Digital Finance, and Inclusion, Volume 1: Cryptocurrency, FinTech, InsurTech, and Regulation*, Cambridge, MA: Academic Press, pp. 37–56.

Newman, H. B., Ellisman, M. H., & Orcutt, J. A. (2003). Data-intensive e-science frontier research. *Communications of the ACM*, **46**(11), 68–77.

Nichols, J. D., Oli, M. K., Kendall, W. L., & Boomer, G. S. (2021). Opinion: A better approach for dealing with reproducibility and replicability in science. *Proceedings of the National Academy of Sciences*, **118**(7), e2100769118.

Nieuwesteeg, B., & Faure, M. (2018). An analysis of the effectiveness of the EU data breach notification obligation. *Computer Law & Security Review*, **34**(6), 1232–1246.

Networking and Information Technology Research and Development (NITRD). (2016). *The Federal Big Data Research and Development Strategic Plan*, Washington, DC: National Science and Technology Council.

Noble, P., van Riper, D., Ruggles, S., Schroeder, J., & Hindman, M. (2011). Harmonizing disparate data across time and place: The Integrated Spatio-Temporal Aggregate Data Series. *Historical Methods*, **44**(2), 79–85.

Nochta, T., Wan, L., Schooling, J. M., & Parlikad, A. K. (2021). A socio-technical perspective on urban analytics: The case of city-scale digital twins. *Journal of Urban Technology*, **28**(1–2), 263–287.

Nosek, B. A., Alter, G., Banks, G. C. et al. (2015). Promoting an open research culture. *Science*, **348**(6242), 1422–1425.

Noulas, A., Scellato, S., Mascolo, C., & Pontil, M. (2011). An Empirical Study of Geographic User Activity Patterns in Foursquare. In *Proceedings of the International AAAI Conference on Web and Social Media*, vol. 5, Barcelona: Association for the Advancement of Artificial Intelligence, pp. 570–573.

Novick, K. A., Biederman, J. A., Desai, A. R. et al. (2018). The AmeriFlux network: A coalition of the willing. *Agricultural and Forest Meteorology*, **249**, 444–456.

Nowviskie, B. (2015). Digital humanities in the Anthropocene. *Digital Scholarship in the Humanities*, **30**(suppl. 1), i4–i15.

Nugent, J. (2018). iNaturalist. *Science Scope*, **41**(7), 12–13.

Nye, B. D. (2015). Intelligent tutoring systems by and for the developing world: A review of trends and approaches for educational technology in a global context. *International Journal of Artificial Intelligence in Education*, **25**(2), 177–203.

O'Hara, A., Shattuck, R. M., & Goerge, R. M. (2017). Linking federal surveys with administrative data to improve research on families. *The Annals of the American Academy of Political and Social Science*, **669**(1), 63–74.

O'Sullivan, D. (2018). Big Data: Why (Oh Why?) This Computational Social Science? In J. Thatcher, A. Shears, & J. Eckert, eds., *Thinking Big Data in Geography: New Regimes, New Research*, Lincoln: University of Nebraska Press, pp. 21–38.

O'Sullivan, D., & Manson, S. M. (2015). Do physicists have "geography envy"? And what can geographers learn from it? *Annals of the Association of American Geographers*, **105** (4), 704–722.

Oliver, J. C., & McNeil, T. (2021). Undergraduate data science degrees emphasize computer science and statistics but fall short in ethics training and domain-specific context. *PeerJ Computer Science*, **7**, e441.

Omrani, H. (2015). Predicting travel mode of individuals by machine learning. *Transportation Research Procedia*, **10**, 840–849.

Openshaw, S. (1983). *The Modifiable Areal Unit Problem*, Norwich: Geo Books.

Openshaw, S. (1984). Ecological fallacies and the analysis of areal census data. *Environment and Planning A*, **16**(1), 17–31.

Openshaw, S. (1992). Some suggestions concerning the development of artificial intelligence tools for spatial modeling and analysis in GIS. *Annals of Regional Science*, **26**, 35–51.

Organisation for Economic Co-operation and Development (OECD). (2007). *OECD Principles and Guidelines for Access to Research Data from Public Funding*, Paris: OECD Publishing, https://doi.org/10.1787/9789264034020-en-fr.

Organisation for Economic Co-operation and Development (OECD). (2013). *The OECD Privacy Framework*, Paris: OECD Publishing.

Oteros-Rozas, E., Martín-López, B., Fagerholm, N., Bieling, C., & Plieninger, T. (2018). Using social media photos to explore the relation between cultural ecosystem services and landscape features across five European sites. *Ecological Indicators*, **94**, 74–86.

Overeem, A., Leijnse, H., & Uijlenhoet, R. (2016). Two and a half years of country-wide rainfall maps using radio links from commercial cellular telecommunication networks. *Water Resources Research*, **52**(10), 8039–8065.

Overpeck, J. T., Meehl, G. A., Bony, S., & Easterling, D. R. (2011). Climate data challenges in the 21st century. *Science*, **331**(6018), 700–702.

Paganini, M., Petiteville, I., Ward, S. et al. (2018). *Satellite Earth Observations in Support of the Sustainable Development Goals: The CEOS Earth Observation Handbook*, Paris: European Space Agency.

Pahl-Wostl, C., Giupponi, C., Richards, K. et al. (2013). Transition towards a new global change science: Requirements for methodologies, methods, data and knowledge. *Environmental Science & Policy*, **28**, 36–47.

Pang, B., & Lee, L. (2008). Opinion mining and sentiment analysis. *Foundations and Trends in Information Retrieval*, **2**(1–2), 1–135.

Paradiso, J. A., & Starner, T. (2005). Energy scavenging for mobile and wireless electronics. *IEEE Pervasive Computing*, **4**(1), 18–27.

Parashar, M. (2019). *Transforming Science through Cyberinfrastructure*, Arlington, VA: National Science Foundation.

Parenteau, M.-P., & Sawada, M. C. (2011). The modifiable areal unit problem (MAUP) in the relationship between exposure to NO_2 and respiratory health. *International Journal of Health Geographics*, **10**(1), 58.

Parhami, B. (2019). Data Longevity and Compatibility. In S. Sakr & A. Y. Zomaya, eds., *Encyclopedia of Big Data Technologies*, New York: Springer, pp. 559–563.

Parida, T., & Ashok, A. (2021). Consolidating Power in the Name of Progress: Techno-solutionism and Farmer Protests in India. In F. Kaltheuner, ed., *Fake AI*, Manchester: Meatspace Press, pp. 161–169.

Parkinson, C. L., Ward, A., & King, M. D. (2006). *Earth Science Reference Handbook*, Washington, DC: National Aeronautics and Space Administration.

Passi, S., & Barocas, S. (2019). Problem Formulation and Fairness. In *Proceedings of the Conference on Fairness, Accountability, and Transparency*, New York: Association for Computing Machinery, pp. 39–48.

Pearl, J., & Mackenzie, D. (2018). *The Book of Why: The New Science of Cause and Effect*, New York: Basic Books.

Peng, R. D. (2011). Reproducible research in computational science. *Science*, **334**(6060), 1226–1227.

Peng, R. D. (2017). Comment on "50 Years of data science." *Journal of Computational and Graphical Statistics*, **26**(4), 767–768.

Pennington, D. D. (2011). Collaborative, cross-disciplinary learning and co-emergent innovation in eScience teams. *Earth Science Informatics*, **4**(2), 55–68.

Pepper, A. (2020). Glass panels and peepholes: Nonhuman animals and the right to privacy. *Pacific Philosophical Quarterly*, **101**(4), 628–650.

Pérez-Hoyos, A., Rembold, F., Kerdiles, H., & Gallego, J. (2017). Comparison of global land cover datasets for cropland monitoring. *Remote Sensing*, **9**(11), 1118.

Pigliucci, M. (2009). The end of theory in science? *EMBO Reports*, **10**(6), 534–534.

Planthaber, G., Stonebraker, M., & Frew, J. (2012). EarthDB: Scalable Analysis of MODIS Data Using SciDB. In *Proceedings of the 1st ACM SIGSPATIAL International Workshop on Analytics for Big Geospatial Data*, New York: Association for Computing Machinery, pp. 11–19.

Plesser, H. E. (2018). Reproducibility vs. replicability: A brief history of a confused terminology. *Frontiers in Neuroinformatics*, **11**, 76.

Poisot, T., Bruneau, A., Gonzalez, A., Gravel, D., & Peres-Neto, P. (2019). Ecological data should not be so hard to find and reuse. *Trends in Ecology & Evolution*, **34**(6), 494–496.

Poisson, A. C., McCullough, I. M., Cheruvelil, K. S. et al. (2019). Quantifying the contribution of citizen science to broad-scale ecological databases. *Frontiers in Ecology and the Environment*, **18**(1), 19–26.

Poorthuis, A. (2018). How to draw a neighborhood? The potential of big data, regionalization, and community detection for understanding the heterogeneous nature of urban neighborhoods. *Geographical Analysis*, **50**(2), 182–203.

Poovey, M. (1998). *A History of the Modern Fact: Problems of Knowledge in the Sciences of Wealth and Society*, Chicago: University of Chicago Press.

Porter, T. M. (1986). *The Rise of Statistical Thinking, 1820–1900*, Princeton, NJ: Princeton University Press.

Portides, D. (2017). Models and Theories. In L. Magnani & T. Bertolotti, eds., *Springer Handbook of Model-Based Science*, Cham: Springer, pp. 25–48.

Posthumus, S., Sinclair, S., & Poplawski, V. (2018). Digital and environmental humanities: Strong networks, innovative tools, interactive objects. *Resilience: A Journal of the Environmental Humanities*, **5**(2), 156–171.

Prasad, S. K., Aghajarian, D., McDermott, M., et al. (2017). Parallel Processing over Spatial-Temporal Datasets from Geo, Bio, Climate and Social Science Communities: A Research Roadmap. In *2017 IEEE International Congress on Big Data (BigData Congress)*, Santa Clara, CA: Institute of Electrical and Electronics Engineers, pp. 232–250.

Prensky, M. (2009). *H. sapiens digital*: From digital immigrants and digital natives to digital wisdom. *Innovate: Journal of Online Education*, **5**(3), 104264.

Presset, B., Laurenczy, B., Malatesta, D., & Barral, J. (2018). Accuracy of a smartphone pedometer application according to different speeds and mobile phone locations in a laboratory context. *Journal of Exercise Science & Fitness*, **16**(2), 43–48.

Prewitt, K. (2010). What is political interference in federal statistics? *The Annals of the American Academy of Political and Social Science*, **631**(1), 225–238.

Prior, L. (2016). In praise of small N, and of N = 1 in particular. *Critical Public Health*, **26** (2), 115–117.

Pritchett, D. (2008). BASE: An acid alternative: In partitioned databases, trading some consistency for availability can lead to dramatic improvements in scalability. *Queue*, **6** (3), 48–55.

Protopop, I., & Shanoyan, A. (2016). Big data and smallholder farmers: Big data applications in the agri-food supply chain in developing countries. *International Food and Agribusiness Management Review*, **19**(A), 173–190.

Puschmann, C., & Powell, A. (2018). Turning words into consumer preferences: How sentiment analysis is framed in research and the news media. *Social Media + Society*, **4** (3), https://doi.org/10.1177/2056305118797724.

Qin, Z. Z., Sander, M. S., Rai, B. et al. (2019). Using artificial intelligence to read chest radiographs for tuberculosis detection: A multi-site evaluation of the diagnostic accuracy of three deep learning systems. *Scientific Reports*, **9**(1), 15000.

Quinn, J., Frias-Martinez, V., & Subramanian, L. (2014). Computational sustainability and artificial intelligence in the developing world. *AI Magazine*, **35**(3), 36–47.

Rabus, B., Eineder, M., Roth, A., & Bamler, R. (2003). The shuttle radar topography mission: A new class of digital elevation models acquired by spaceborne radar. *ISPRS Journal of Photogrammetry and Remote Sensing*, **57**(4), 241–262.

Rana, A. N. (2020). *Leveraging Big Data to Advance Gender Equality*, Washington, DC: World Bank.

Ray, S., Demke Brown, A., Koudas, N., Blanco, R., & Goel, A. K. (2015). Parallel In-Memory Trajectory-Based Spatiotemporal Topological Join. In *Big Data: 2015 IEEE International Conference on Big Data*, Santa Clara, CA: Institute of Electrical and Electronics Engineers, pp. 361–370.

Reed, D. A., & Dongarra, J. (2015). Exascale computing and big data. *Communications of the ACM*, **58**(7), 56–68.

Reichstein, M., Camps-Valls, G., Stevens, B. et al. (2019). Deep learning and process understanding for data-driven Earth system science. *Nature*, **566**(7743), 195–204.

Reinsel, D., Gantz, J., & Rydning, J. (2018). *Data Age 2025: The Digitization of the World from Edge to Core*, Needham, MA: International Data Corporation (sponsored by Seagate).

Réjou-Méchain, M., Barbier, N., Couteron, P. et al. (2019). Upscaling forest biomass from field to satellite measurements: Sources of errors and ways to reduce them. *Surveys in Geophysics*, **40**(4), 881–911.

Retchless, D. (2018). Bringing the Big Data of Climate Change Down to Human Scale: Citizen Sensors and Personalized Visualizations in Climate Communication. In J. Thatcher, A. Shears, & J. Eckert, eds., *Thinking Big Data in Geography: New Regimes, New Research*, Lincoln: University of Nebraska Press, pp. 197–213.

Rieppel, O. (2010). New essentialism in biology. *Philosophy of Science*, **77**(5), 662–673.

Robinson, C., & Franklin, R. S. (2021). The sensor desert quandary: What does it mean (not) to count in the smart city? *Transactions of the Institute of British Geographers*, **46** (2), 238–254.

Robinson, C., Malkin, K., Jojic, N. et al. (2021). Global Land-cover mapping with weak supervision: Outcome of the 2020 IEEE GRSS Data Fusion Contest. *IEEE Journal of Selected Topics in Applied Earth Observations and Remote Sensing*, **14**, 3185–3199.

Rohde, R. A., & Hausfather, Z. (2020). The Berkeley Earth Land/Ocean Temperature Record. *Earth System Science Data*, **12**(4), 3469–3479.

Román-Rivera, M. A., & Ellis, J. T. (2019). A synthetic review of remote sensing applications to detect nearshore bars. *Marine Geology*, **408**, 144–153.

Rosa, E. A., Diekmann, A., Dietz, T., & Jaeger, C. (2010). *Human Footprints on the Global Environment: Threats to Sustainability*, Cambridge, MA: MIT Press.

Rosenberg, D. (2013). Data before the Fact. In L. Gitelman, ed., *Raw Data Is an Oxymoron*, Cambridge, MA: MIT Press, pp. 15–40.

Ross, C. J., Wolfe, N., Plagge, M. et al. (2019). Using Scientific Visualization Techniques to Visualize Parallel Network Simulations. In *Proceedings of the 2019 ACM SIGSIM Conference on Principles of Advanced Discrete Simulation*, New York: Association for Computing Machinery, pp. 197–200.

Rowley, J. (2007). The wisdom hierarchy: Representations of the DIKW hierarchy. *Journal of Information Science*, **33**(2), 163–180.

Ruddiman, W. F. (2013). The Anthropocene. *Annual Review of Earth and Planetary Sciences*, **41**, 45–68.

Ruggles, S. (2014). Big microdata for population research. *Demography*, **51**(1), 287–297.

Ruggles, S., & van Riper, D. (2022). The role of chance in the census bureau database reconstruction experiment. *Population Research and Policy Review*, **41**, 781–788.

Runck, B. C., Manson, S., Shook, E., Gini, M., & Jordan, N. (2019). Using word embeddings to generate data-driven human agent decision-making from natural language. *GeoInformatica*, **23**(2), 221–242.

Runge, J. (2018). Causal network reconstruction from time series: From theoretical assumptions to practical estimation. *Chaos: An Interdisciplinary Journal of Nonlinear Science*, **28**(7), 75310.

Ruppert, E. (2013). Rethinking empirical social sciences. *Dialogues in Human Geography*, **3**(3), 268–273.

Rzeszewski, M. (2018). Geosocial capta in geographical research: A critical analysis. *Cartography and Geographic Information Science*, **45**(1), 18–30.

Sachs, J. D. (2012). From Millennium Development Goals to Sustainable Development Goals. *The Lancet*, **379**(9832), 2206–2211.

Sadowski, J. (2021). "Anyway, the dashboard is dead": On trying to build urban informatics. *New Media & Society*, **24**(forthcoming), https://doi.org/10.1177/14614448211058455.

Salganik, M. J., Lundberg, I., Kindel, A. T. et al. (2020). Measuring the predictability of life outcomes with a scientific mass collaboration. *Proceedings of the National Academy of Sciences*, **117**(15), 8398–8403.

Salmond, J. A., Tadaki, M., & Dickson, M. (2017). Can big data tame a "naughty" world? *Canadian Geographer*, **61**(1), 52–63.

Samek, W., Montavon, G., Lapuschkin, S., Anders, C. J., & Müller, K.-R. (2021). Explaining deep neural networks and beyond: A review of methods and applications. *Proceedings of the IEEE*, **109**(3), 247–278.

Sánchez-Clavijo, L., Martinez, S., Acevedo-Charry, O. et al. (2021). Differential reporting of biodiversity in two citizen science platforms during COVID-19 lockdown in Colombia. *Biological Conservation*, **256**, 109077.

Sandefur, J., & Glassman, A. (2015). The political economy of bad data: Evidence from African survey and administrative statistics. *The Journal of Development Studies*, **51**(2), 116–132.

Sanderson, J., Wiseman, L., & Poncini, S. (2018). What's behind the ag-data logo? An examination of voluntary agricultural-data codes of practice. *International Journal of Rural Law and Policy*, **1**(1), 6043.

Sarhadi, A., Burn, D. H., Yang, G., & Ghodsi, A. (2017). Advances in projection of climate change impacts using supervised nonlinear dimensionality reduction techniques. *Climate Dynamics*, **48**(3–4), 1329–1351.

Savage, M., & Burrows, R. (2007). The coming crisis of empirical sociology. *Sociology*, **41** (5), 885–899.

Sawyer, S. (2008). Data wealth, data poverty, science and cyberinfrastructure. *Prometheus*, **26**(4), 355–371.

Schepaschenko, D., See, L., Lesiv, M. et al. (2015). Development of a global hybrid forest mask through the synergy of remote sensing, crowdsourcing and FAO statistics. *Remote Sensing of Environment*, **162**, 208–220.

Scher, S. (2018). Toward data-driven weather and climate forecasting: Approximating a simple general circulation model with deep learning. *Geophysical Research Letters*, **45**(22), 12, 616–12, 622.

Schneble, C. O., Elger, B. S., & Shaw, D. (2018). The Cambridge Analytica affair and internet-mediated research. *EMBO Reports*, **19**(8), e46579.

Schöpfel, J., & Azeroual, O. (2021). Rewarding Research Data Management. In *Companion Proceedings of the Web Conference 2021*, New York: Association for Computing Machinery, pp. 446–450.

Schrodt, F., Kattge, J., Shan, H. et al. (2015). BHPMF: A hierarchical Bayesian approach to gap-filling and trait prediction for macroecology and functional biogeography. *Global Ecology and Biogeography*, **24**(12), 1510–1521.

Sculley, D., Snoek, J., Wiltschko, A., & Rahimi, A. (2018). Winner's Curse? On Pace, Progress, and Empirical Rigor. In *ICLR 2018 Workshop*, Vancouver: International Conference on Learning Representations. Accessed September 2, 2022, https://openre view.net/forum?id=rJWFOFywf.

Sefara, T. J., Mokgonyane, T. B., Manamela, M. J., & Modipa, T. I. (2019). HMM-Based Speech Synthesis System Incorporated with Language Identification for Low-Resourced Languages. In *2019 International Conference on Advances in Big Data, Computing and Data Communication Systems (icABCD)*, Santa Clara, CA: Institute of Electrical and Electronics Engineers.

Sejnowski, T. J. (2020). The unreasonable effectiveness of deep learning in artificial intelligence. *Proceedings of the National Academy of Sciences*, **117**(48), 30033–30038.

Selbst, A. D. (2021). An institutional view of algorithmic impact assessments. *Harvard Journal of Law & Technology*, **35**(1), 117–191.

Selinger, E., & Hartzog, W. (2015). Facebook's emotional contagion study and the ethical problem of co-opted identity in mediated environments where users lack control. *Research Ethics*, **12**(1), 35–43.

Serra, Y. L., Haase, J. S., Adams, D. K. et al. (2018). The risks of contracting the acquisition and processing of the nation's weather and climate data to the private sector. *Bulletin of the American Meteorological Society*, **99**(5), 869–870.

Shearmur, R. (2010). Editorial – A world without data? The unintended consequences of fashion in geography. *Urban Geography*, **31**(8), 1009–1017.

Shearmur, R. (2015). Dazzled by data: Big Data, the census and urban geography. *Urban Geography*, **36**(7), 965–968.

Sheehy, C. J. (2019). Christine Baeumler's environmental art: A pollinator garden as a life practice. *Public Art Dialogue*, **9**(2), 193–217.

Shetal, A., Feng, Z., & Savani, K. (2020). Using machine learning to generate novel hypotheses: Increasing optimism about COVID-19 makes people less willing to justify unethical behaviors. *Psychological Science*, **31**(10), 1222–1235.

Shook, E., Hodgson, M. E., Wang, S. et al. (2016). Parallel cartographic modeling: A methodology for parallelizing spatial data processing. *International Journal of Geographical Information Science*, **30**(12), 2355–2376.

Shook, E., & Turner, V. K. (2016). The socio-environmental data explorer (SEDE): A social media–enhanced decision support system to explore risk perception to hazard events. *Cartography and Geographic Information Science*, **43**(5), 427–441.

Shoval, N., Kwan, M.-P., Reinau, K. H., & Harder, H. (2014). The shoemaker's son always goes barefoot: Implementations of GPS and other tracking technologies for geographic research. *Geoforum*, **51**, 1–5.

Sieber, R., & Tenney, M. (2018). Smaller and Slower Data in an Era of Big Data. InJ. Thatcher, A. Shears, & J. Eckert, eds., *Thinking Big Data in Geography: New Regimes, New Research*, Lincoln: University of Nebraska Press, pp. 41–69.

Silver, D., Schrittwieser, J., Simonyan, K. et al. (2017). Mastering the game of go without human knowledge. *Nature*, **550**(7676), 354–359.

Simmons, J. P., Nelson, L. D., & Simonsohn, U. (2011). False-positive psychology: Undisclosed flexibility in data collection and analysis allows presenting anything as significant. *Psychological Science*, **22**(11), 1359–1366.

Singh, D., & Reddy, C. K. (2014). A survey on platforms for big data analytics. *Journal of Big Data*, **2**(1), 8.

Singleton, A., Arribas-Bel, D., Murray, J., & Fleischmann, M. (2022). Estimating generalized measures of local neighbourhood context from multispectral satellite images using a convolutional neural network. *Computers, Environment and Urban Systems*, **95**(2022), 101802.

Slaughter, M. J., & McCormick, D. H. (2021). Data Is Power Washington Needs to Craft New Rules for the Digital Age. *Foreign Affairs*, May/June, 54–60.

Smith, B. H. (2016). What was "close reading"? A century of method in literary studies. *The Minnesota Review*, 2016(87), 57–75.

Smith, H., Medero, G. M., Crane De Narváez, S., & Castro Mera, W. (2022). Exploring the relevance of "smart city" approaches to low-income communities in Medellín, Colombia. *GeoJournal*, **87**(forthcoming), http://dx.doi.org/10.1007/s10708-022-10574-y.

Smith, L. M., Barth, J. A., Kelley, D. S. et al. (2018). The ocean observatories initiative. *Oceanography*, **31**(1), 16–35.

Smith, M. W., Carrivick, J. L., & Quincey, D. J. (2016). Structure from motion photogrammetry in physical geography. *Progress in Physical Geography*, **40**(2), 247–275.

Smith, N. R. (2000). The global ocean data assimilation experiment. *Advances in Space Research*, **25**(5), 1089–1098.

Snyder, J. P. (1987). *Map Projections: A Working Manual*, Washington, DC: United States Geological Survey.

Sobek, M., Cleveland, L., Flood, S. et al. (2011). Big data: Large-scale historical infrastructure from the Minnesota Population Center. *Historical Methods*, **44**(2), 61–68.

Solove, D. J. (2008). *Understanding Privacy*, Cambridge, MA: Harvard University Press.

Song, X.-P., Huang, C., & Townshend, J. R. (2017). Improving global land cover characterization through data fusion. *Geo-Spatial Information Science*, **20**(2), 141–150.

Sood, S. K., Sandhu, R., Singla, K., & Chang, V. (2018). IoT, big data and HPC based smart flood management framework. *Sustainable Computing: Informatics and Systems*, **20**, 102–117.

Spasser, M. A. (2000). Articulating collaborative activity: Design-in-use of collaborative publishing services in the Flora of North America Project. *Scandinavian Journal of Information Systems*, **12**(1), 149–172.

Spies, T. A., White, E., Ager, A. et al. (2017). Using an agent-based model to examine forest management outcomes in a fire-prone landscape in Oregon, USA. *Ecology and Society*, **22**(1), 25.

Stafford, R., Hart, A. G., Collins, L. et al. (2010). Eu-social science: The role of Internet social networks in the collection of bee biodiversity data. *PLOS ONE*, **5**(12), e14381.

Star, S. L., & Ruhleder, K. (1996). Steps toward an ecology of infrastructure: Design and access for large information spaces. *Information Systems Research*, **7**(1), 111–134.

Steele, J. E., Sundsøy, P. R., Pezzulo, C. et al. (2017). Mapping poverty using mobile phone and satellite data. *Journal of the Royal Society Interface*, **14**(127), 20160690.

Stern, P. C. (1993). A second environmental science: Human–environment interactions. *Science*, **260**, 1897–1899.

Stewart, C. A., Roskies, R., Knepper, R. et al. (2015). XSEDE Value Added, Cost Avoidance, and Return on Investment. In *Proceedings of the 2015 XSEDE Conference: Scientific Advancements Enabled by Enhanced Cyberinfrastructure*, St. Louis, MO: Association for Computing Machinery, p. 23.

Stocks, K. I., Schramski, S., Virapongse, A., & Kempler, L. (2019). Geoscientists' perspectives on cyberinfrastructure needs: A collection of user scenarios. *Data Science Journal*, **18**(1), 21.

Stodden, V., McNutt, M., Bailey, D. H. et al. (2016). Enhancing reproducibility for computational methods. *Science*, **354**(6317), 1240–1241.

Stodden, V., Seiler, J., & Ma, Z. (2018). An empirical analysis of journal policy effectiveness for computational reproducibility. *Proceedings of the National Academy of Sciences*, **115**(11), 2584–2589.

Stone, Z. (2019). Under the Influence of a "Super Bloom." New York Times. May 23. Accessed August 16, 2022, www.nytimes.com/2019/03/23/style/super-bloom-california -instagram-influencer.html.

Strasser, B. J. (2012). Data-driven sciences: From wonder cabinets to electronic databases. *Studies in History and Philosophy of Science Part C: Studies in History and Philosophy of Biological and Biomedical Sciences*, **43**(1), 85–87.

Struijs, P., Braaksma, B., & Daas, P. J. (2014). Official statistics and big data. *Big Data & Society*, **1**(1), https://doi.org/10.1177/2053951714538417.

Stump, C. (2021). Artificial intelligence aids intuition in mathematical discovery. *Nature*, **600**, 44–45.

Sudmanns, M., Tiede, D., Lang, S. et al. (2019). Big Earth data: disruptive changes in Earth observation data management and analysis? *International Journal of Digital Earth*, **13**(7), 832–850.

Suh, Y.-K., & Lee, K. Y. (2018). A survey of simulation provenance systems: Modeling, capturing, querying, visualization, and advanced utilization. *Human-Centric Computing and Information Sciences*, **8**(1), 27.

Suhr, B., Dungl, J., & Stocker, A. (2020). Search, reuse and sharing of research data in materials science and engineering: A qualitative interview study. *PLOS ONE*, **15**(9), e0239216.

Sullivan, B. L., Wood, C. L., Iliff, M. J. et al. (2009). eBird: A citizen-based bird observation network in the biological sciences. *Biological Conservation*, **142**(10), 2282–2292.

Sundberg, J. (2013). Decolonizing posthumanist geographies. *Cultural Geographies*, **21**(1), 33–47.

Sutton, R. S., & Barto, A. G. (2018). *Reinforcement Learning: An Introduction*, Cambridge, MA: MIT press.

Swain, R. B. (2018). A Critical Analysis of the Sustainable Development Goals. In W. Leal Filho, ed., *Handbook of Sustainability Science and Research*, New York: Springer, pp. 341–355.

Sward, D., Craig, W., Delegard, K. et al. (2022). Spatial University for Service and Support. In S. M. Manson, L. Kne, B. Krzyzanowski, & J. Lindelof, eds., *Building the Spatial University*, Cham: Springer, pp. 31–54.

Taddeo, M., & Floridi, L. (2018). How AI can be a force for good. *Science*, **361**(6404), 751–752.

Tarboton, D. G., Idaszak, R., Horsburgh, J. S. et al. (2014). Hydro Share: Advancing Collaboration through Hydrologic Data and Model Sharing. In *7th International Congress on Environmental Modelling and Software (iEMSs 2014)*, San Diego, CA: International Environmental Modelling and Software Society, http://dx.doi.org/10.13140/2.1.4431.6801.

Tatem, A. J., Qiu, Y., Smith, D. L. et al. (2009). The use of mobile phone data for the estimation of the travel patterns and imported *Plasmodium falciparum* rates among Zanzibar residents. *Malaria Journal*, **8**(1), 287.

Taylor, L., & Schroeder, R. (2015). Is bigger better? The emergence of big data as a tool for international development policy. *GeoJournal*, **80**(4), 503–518.

Telikani, A., Gandomi, A. H., & Shahbahrami, A. (2020). A survey of evolutionary computation for association rule mining. *Information Sciences*, **524**, 318–352.

That, D. H. T., Fils, G., Yuan, Z., & Malik, T. (2017). Sciunits: Reusable Research Objects. In *2017 IEEE 13th International Conference on e-Science (e-Science)*, Santa Clara, CA: Institute of Electrical and Electronics Engineers, pp. 374–383.

Thatcher, J., O'Sullivan, D., & Mahmoudi, D. (2016). Data colonialism through accumulation by dispossession: New metaphors for daily data. *Environment and Planning D: Society and Space*, **34**(6), 990–1006.

Thenkabail, P. S. (2019). Remote Sensing Data Characterization, Classification, and Accuracies: Advances of the Last 50 Years and a Vision for the Future. In *Remotely Sensed Data Characterization, Classification, and Accuracies*, Boca Raton, FL: CRC Press, pp. 659–696.

Thinyane, M., Goldkind, L., & Lam, H. I. (2018). Data collaboration and participation for Sustainable Development Goals: A case for engaging community-based organizations. *Journal of Human Rights and Social Work*, **3**(1), 44–51.

Thurgate, N., Lowe, A. J., Clancy, T. F., Chabbi, A., & Loescher, H. W. (2017). Australia's Terrestrial Ecosystem Research Network: A Network of Networks Approach to Building and Maintaining Continental Ecosystem Research Infrastructures. In A. Chabbi & H. W. Loescher, eds., *Terrestrial Ecosystem Research Infrastructures: Challenges, New Developments and Perspectives*, London: Routledge, pp. 427–448.

Torney, C. J., Lloyd-Jones, D. J., Chevallier, M. et al. (2019). A comparison of deep learning and citizen science techniques for counting wildlife in aerial survey images. *Methods in Ecology and Evolution*, **10**(6), 779–787.

Touzé-Peiffer, L., Barberousse, A., & Le Treut, H. (2020). The Coupled Model Intercomparison Project: History, uses, and structural effects on climate research. *WIREs Climate Change*, **11**(4), e648.

Towns, J. (2018). Toward an open, sustainable national advanced computing ecosystem. *Computing in Science & Engineering*, **20**(5), 39–46.

Towns, J., Cockerill, T., Dahan, M. et al. (2014). XSEDE: Accelerating scientific discovery. *Computing in Science & Engineering*, **16**(5), 62–74.

Townsend, A. M. (2013). *Smart Cities: Big Data, Civic Hackers, and the Quest for a New Utopia*, New York: WW Norton & Company.

Toyama, K. (2010). Can technology end poverty? *Boston Review*, **36**(5), 12–29.

Travis, C., & Holm, P. (2016). The Digital Environmental Humanities: What Is It and Why Do We Need It? The NorFish Project and SmartCity Lifeworlds. In *The Digital Arts and Humanities*, Cham: Springer, pp. 187–204.

Trethewie, S. (2021). New Norms and New Challenges in Food Security. In A. D. B. Cook & T. Nair, eds., *Non-Traditional Security in the Asia-Pacific*, Singapore: World Scientific, pp. 91–94.

Trilles, S., Belmonte, Ò. , Schade, S., & Huerta, J. (2017). A domain-independent methodology to analyze IoT data streams in real-time: A proof of concept implementation for anomaly detection from environmental data. *International Journal of Digital Earth*, **10** (1), 103–120.

Tsonis, A. A., Swanson, K. L., & Roebber, P. J. (2006). What do networks have to do with climate? *Bulletin of the American Meteorological Society*, **87**(5), 585–596.

Tukey, J. W. (1962). The Future of Data Analysis. *Annals of Mathematical Statistics*, **33** (1), 1–67.

Turing, A. M. (2009). Computing Machinery and Intelligence. In R. Epstein, G. Roberts, & G. Beber, eds., *Parsing the Turing Test: Philosophical and Methodological Issues in the Quest for the Thinking Computer*, Cham: Springer, pp. 23–65.

Turnhout, E., & Lahsen, M. (2022). Transforming environmental research to avoid tragedy. *Climate and Development*, **14**(forthcoming), https://doi.org/10.1080/17565529 .2022.2062287.

Uhrqvist, O., & Linnér, B.-O. (2015). Narratives of the past for Future Earth: The historiography of global environmental change research. *The Anthropocene Review*, **2**(2), 159–173.

United Nations. (2014). *Fundamental Principles of Official Statistics*, New York: United Nations.

United Nations. (2020). *Sustainable Development Goals: Guidelines for the Use of the SDG Logo Including the Colour Wheel, and 17 Icons*, New York: United Nations Department of Global Communications.

van der Wal, R., Zeng, C., Heptinstall, D. et al. (2015). Automated data analysis to rapidly derive and communicate ecological insights from satellite-tag data: A case study of reintroduced red kites. *Ambio*, **44**(4), 612–623.

van Ginkel, M., & Biradar, C. (2021). Drought early warning in agri-food systems. *Climate*, **9**(9), 134.

van Meter, H. J. (2020). Revising the DIKW pyramid and the real relationship between data, information, knowledge and wisdom. *Law, Technology and Humans*, **2**(2), 69–80.

van Schaik, P., Jansen, J., Onibokun, J., Camp, J., & Kusev, P. (2018). Security and privacy in online social networking: Risk perceptions and precautionary behaviour. *Computers in Human Behavior*, **78**, 283–297.

VanValkenburgh, P., & Dufton, J. A. (2020). Big archaeology: Horizons and blindspots. *Journal of Field Archaeology*, **45**(suppl. 1), S1–S7.

Vardi, M. Y. (2022). ACM, ethics, and corporate behavior. *Communications of the ACM*, **65** (3), 5.

Vardigan, M., Heus, P., & Thomas, W. (2008). Data documentation initiative: Toward a standard for the social sciences. *International Journal of Digital Curation*, **3**(1), 107–113.

Veale, M., & Binns, R. (2017). Fairer machine learning in the real world: Mitigating discrimination without collecting sensitive data. *Big Data & Society*, **4**(2), https://doi .org/10.1177/2053951717743530.

Venkatramanan, S., Sadilek, A., Fadikar, A. et al. (2021). Forecasting influenza activity using machine-learned mobility map. *Nature Communications*, **12**, 726.

Vinuesa, R., Azizpour, H., Leite, I. et al. (2020). The role of artificial intelligence in achieving the Sustainable Development Goals. *Nature Communications*, **11**, 233.

Virdee, S. (2019). Racialized capitalism: An account of its contested origins and consolidation. *The Sociological Review*, **67**(1), 3–27.

Vitak, J., Proferes, N., Shilton, K., & Ashktorab, Z. (2017). Ethics regulation in social computing research: Examining the role of institutional review boards. *Journal of Empirical Research on Human Research Ethics*, **12**(5), 372–382.

Vitak, J., Shilton, K., & Ashktorab, Z. (2016). Beyond the Belmont Principles. In *Proceedings of the 19th ACM Conference on Computer-Supported Cooperative Work & Social Computing*, New York: Association for Computing Machinery, pp. 941–953.

Vitolo, C., Elkhatib, Y., Reusser, D., Macleod, C. J. A., & Buytaert, W. (2015). Web technologies for environmental big data. *Environmental Modelling & Software*, **63**, 185–198.

Voinov, A., & Bousquet, F. (2010). Modelling with stakeholders. *Environmental Modelling & Software*, **25**(11), 1268–1281.

Voosen, P. (2017). The AI detectives. *Science*, **357**(6346), 22–27.

VoPham, T., Hart, J. E., Laden, F., & Chiang, Y.-Y. (2018). Emerging trends in geospatial artificial intelligence (geoAI): Potential applications for environmental epidemiology. *Environmental Health*, **17**, 40.

Walford, A. (2018). "If Everything Is Information": Archives and Collecting on the Frontiers of Data-Driven Science. In H. Knox & D. Nafus, eds., *Ethnography for a Data-Saturated World*, Manchester: Manchester University Press, pp. 105–127.

Waliser, D., Gleckler, P. J., Ferraro, R. et al. (2020). Observations for Model Intercomparison Project (Obs4MIPs): Status for CMIP6. *Geoscientific Model Development*, **13**(7), 2945–2958.

Walker, M. A. (2015). The professionalisation of data science. *International Journal of Data Science*, **1**(1), 7–16.

Wallace, T. R., Watkins, D., & Schwartz, J. (2018). Where We Live: A Map of Every Building in America. *New York Times*. October 12. Accessed August 16, 2022, www.nytimes.com /interactive/2018/10/12/us/map-of-every-building-in-the-united-states.html.

Waller, M. A., & Fawcett, S. E. (2013). Click here for a data scientist: Big data, predictive analytics, and theory development in the era of a maker movement supply chain. *Journal of Business Logistics*, **34**(4), 249–252.

Wang, C., Chen, M.-H., Schifano, E., Wu, J., & Yan, J. (2016). Statistical methods and computing for big data. *Statistics and Its Interface*, **9**(4), 399–414.

Wang, S. (2010). A CyberGIS framework for the synthesis of cyberinfrastructure, GIS, and spatial analysis. *Annals of the Association of American Geographers*, **100**(3), 535–557.

Wang, S. (2016). CyberGIS and spatial data science. *GeoJournal*, **81**(6), 965–968.

Wang, V., & Shepherd, D. (2020). Exploring the extent of openness of open government data: A critique of open government datasets in the UK. *Government Information Quarterly*, **37**(1), 101405.

Wang, Y., & Allen, T. R. (2008). Estuarine shoreline change detection using Japanese ALOS PALSAR HH and JERS-1 L-HH SAR data in the Albemarle-Pamlico Sounds, North Carolina, USA. *International Journal of Remote Sensing*, **29**(15), 4429–4442.

Waranch, R. S. (2017). Digital rights Ireland deja vu: Why the bulk acquisition warrant provisions of the Investigatory Powers Act 2016 are incompatible with the Charter of Fundamental Rights of the European Union. *The George Washington International Law Review*, **50**, 209–242.

Warf, B. (2013). Contemporary digital divides in the United States. *Tijdschrift Voor Economische En Sociale Geografie*, **104**(1), 1–17.

Warf, B., & Sui, D. (2010). From GIS to neogeography: Ontological implications and theories of truth. *Annals of GIS*, **16**(4), 197–209.

Warneke, B., Last, M., Liebowitz, B., & Pister, K. S. J. (2001). Smart dust: Communicating with a cubic-millimeter computer. *Computer*, **34**(1), 44–51.

Watts, D. J. (2017). Should social science be more solution-oriented? *Nature Human Behaviour*, **1**, 15.

Watts, V. (2013). Indigenous place-thought and agency amongst humans and non humans (First Woman and Sky Woman go on a European world tour!). *Decolonization: Indigeneity, Education & Society*, **2**(1), 20–34.

Weidemann, C. D., Swift, J. N., & Kemp, K. K. (2018). Geosocial Footprints and Geoprivacy Concerns. In J. Thatcher, A. Shears, & J. Eckert, eds., *Thinking Big Data in Geography: New Regimes, New Research*, Lincoln: University of Nebraska Press, pp. 123–144.

Weinberger, D. (2011). *Too Big to Know: Rethinking Knowledge Now That the Facts Aren't the Facts, Experts Are Everywhere, and the Smartest Person in the Room Is the Room*, New York: Basic Books.

Weiss, K., Khoshgoftaar, T. M., & Wang, D. (2016). A survey of transfer learning. *Journal of Big Data*, **3**(1), 1–40.

Weiss, S. M., & Indurkhya, N. (1998). *Predictive Data Mining: A Practical Guide*, San Francisco: Morgan Kaufmann Publishers.

Wesolowski, A., Eagle, N., Noor, A. M., Snow, R. W., & Buckee, C. O. (2013). The impact of biases in mobile phone ownership on estimates of human mobility. *Journal of the Royal Society Interface*, **10**(81), 20120986.

Whiley, A. (2018). *Data Strategy 2018–20: Valuing and Using Trusted Data and Analytics in Our Decisions*, Canberra: Department of Industry, Innovation and Science.

White, P., & Breckenridge, R. S. (2014). Trade-offs, limitations, and promises of big data in social science research. *Review of Policy Research*, **31**(4), 331–338.

Wickham, H., & Grolemund, G. (2016). *R for Data Science: Import, Tidy, Transform, Visualize, and Model Data*, Sebastopol, CA: O'Reilly Media, Inc.

Wieczorek, J., Bloom, D., Guralnick, R. et al. (2012). Darwin Core: An evolving community-developed biodiversity data standard. *PLOS ONE*, **7**(1), e29715.

Wilkins-Diehr, N., Gannon, D., Klimeck, G., Oster, S., & Pamidighantam, S. (2008). TeraGrid science gateways and their impact on science. *Computer*, **41**(11), 32–41.

Wilkinson, G. G. (1996). A review of current issues in the integration of GIS and remote sensing data. *International Journal of Geographical Information Systems*, **10**(1), 85–101.

Wilkinson, J., Scott, C. J., & Willis, D. M. (2016a). Going with the floe. *Astronomy & Geophysics*, **57**(2), 2–37.

Wilkinson, M. D., Dumontier, M., Aalbersberg, I J. J. et al. (2016b). The FAIR Guiding principles for scientific data management and stewardship. *Scientific Data*, **3**(1), 160018.

Willcock, S., Martínez-López, J., Hooftman, D. A. P. et al. (2018). Machine learning for ecosystem services. *Ecosystem Services*, **33**, 165–174.

Williams, J. H., DeBenedictis, A., Ghanadan, R. et al. (2012). The technology path to deep greenhouse gas emissions cuts by 2050: The pivotal role of electricity. *Science*, **335**(6064), 53–59.

Williams, S. (2020). *Data Action: Using Data for Public Good*. Cambridge, MA: MIT Press.

Wiseman, L., Sanderson, J., Zhang, A., & Jakku, E. (2019). Farmers and their data: An examination of farmers' reluctance to share their data through the lens of the laws impacting smart farming. *NJAS – Wageningen Journal of Life Sciences*, 90–91, 100301.

Wood, S. A., Guerry, A. D., Silver, J. M., & Lacayo, M. (2013). Using social media to quantify nature-based tourism and recreation. *Scientific Reports*, **3**(1), 2976.

Wooldridge, M. (2021). *A Brief History of Artificial Intelligence: What It Is, Where We Are, and Where We Are Going*, New York: Flatiron Books.

World Commission on Environment and Development. (1987). *Our Common Future*, Oxford: United Nations and Oxford University Press.

Wright, D., Gutwirth, S., Friedewald, M., Vildjiounaite, E., & Punie, Y. (2008). *Safeguards in a World of Ambient Intelligence*, Berlin: Springer.

Wright, L. G., Onodera, T., Stein, M. M. et al. (2022). Deep physical neural networks trained with backpropagation. *Nature*, **601**(7894), 549–555.

Wu, C., Zhu, Q., Zhang, Y. et al. (2017). A NoSQL–SQL hybrid organization and management approach for real-time geospatial data: A case study of public security video surveillance. *ISPRS International Journal of Geo-Information*, **6** (1), 21.

Wüest, R. O., Zimmermann, N. E., Zurell, D. et al. (2020). Macroecology in the age of big data: Where to go from here? *Journal of Biogeography*, **47**(1), 1–12.

Wulder, M. A., Coops, N. C., Roy, D. P., White, J. C., & Hermosilla, T. (2018). Land cover 2.0. *International Journal of Remote Sensing*, **39**(12), 4254–4284.

Wyly, E. (2014). Automated (post)positivism. *Urban Geography*, **35**(5), 669–690.

Xie, K., Yang, D., Ozbay, K., & Yang, H. (2019). Use of real-world connected vehicle data in identifying high-risk locations based on a new surrogate safety measure. *Accident Analysis & Prevention*, **125**, 311–319.

Xie, Y., Cai, J., Bhojwani, R., Shekhar, S., & Knight, J. (2020). A locally-constrained yolo framework for detecting small and densely-distributed building footprints. *International Journal of Geographical Information Science*, **34**(4), 777–801.

Xu, H., Russell, T., Coposky, J. et al. (2017). iRODS primer 2: Integrated rule-oriented data system. *Synthesis Lectures on Information Concepts, Retrieval, and Services*, **9**(3), 1–131.

Yang, C., & Huang, Q. (2013). *Spatial Cloud Computing: A Practical Approach*, Boca Raton, FL: CRC Press.

Yang, C., Huang, Q., Li, Z., Liu, K., & Hu, F. (2017a). Big data and cloud computing: innovation opportunities and challenges. *International Journal of Digital Earth*, **10**(1), 13–53.

Yang, C., Yu, M., Hu, F., Jiang, Y., & Li, Y. (2017b). Utilizing cloud computing to address big geospatial data challenges. *Computers, Environment and Urban Systems*, **61**, 120–128.

Yang, L., & Gilbert, N. (2008). Getting away from numbers: Using qualitative observation for agent-based modeling. *Advances in Complex Systems*, **11**(2), 175–185.

Yin, D., Liu, Y., Hu, H. et al. (2019). CyberGIS-Jupyter for reproducible and scalable geospatial analytics. *Concurrency and Computation: Practice and Experience*, **31**(11), e5040.

Zaharia, M., Chowdhury, M., Das, T. et al. (2012). Resilient Distributed Datasets: A Fault-Tolerant Abstraction for In-Memory Cluster Computing. In *Proceedings of the 9th USENIX Conference on Networked Systems Design and Implementation*, San Jose, CA: USENIX Association, pp. 15–28.

Zegura, E., DiSalvo, C., & Meng, A. (2018). Care and the Practice of Data Science for Social Good. In *Proceedings of the 1st ACM SIGCAS Conference on Computing and Sustainable Societies*, New York: Association for Computing Machinery.

Zeng, W., Lin, C., Lin, J. et al. (2020). Revisiting the modifiable areal unit problem in deep traffic prediction with visual analytics. *IEEE Transactions on Visualization and Computer Graphics*, **27**(2), 839–848.

Zhang, M., Alvarez, R. M., & Levin, I. (2019). Election forensics: Using machine learning and synthetic data for possible election anomaly detection. *PLOS ONE*, **14**(10), e0223950.

Zhao, L., Song, C. X., Kalyanam, R. et al. (2017). GABBs: Reusable Geospatial Data Analysis Building Blocks for Science Gateways. In *Ninth International Workshop on Science Gateways*, Poznan: International Workshop on Science Gateways.

Zhao, Z., & Hellström, M. (2020). *Towards Interoperable Research Infrastructures for Environmental and Earth Sciences: A Reference Model Guided Approach for Common Challenges*, Cham: Springer.

Zheng, F., Tao, R., Maier, H. R. et al. (2018). Crowdsourcing methods for data collection in geophysics: State of the art, issues, and future directions. *Reviews of Geophysics*, **56**(4), 698–740.

Zhou, Y., Guo, Q., Sun, H. et al. (2019). A novel data-driven approach for transient stability prediction of power systems considering the operational variability. *International Journal of Electrical Power & Energy Systems*, **107**, 379–394.

Žliobaitė, I., & Custers, B. (2016). Using sensitive personal data may be necessary for avoiding discrimination in data-driven decision models. *Artificial Intelligence and Law*, **24**(2), 183–201.

Zomaya, A. Y., & Sakr, S. (2017). *Handbook of Big Data Technologies*, Cham: Springer.

Zuboff, S. (2019). *The Age of Surveillance Capitalism: The Fight for a Human Future at the New Frontier of Power*, New York: Public Affairs.

Index

abductive reasoning. *See* epistemology: abductive reasoning

accessibility. *See* FAIR principles

accuracy, 8, 11, 27, 30, 35, 59, 61

active data, 34

administrative data, 40, 43, 68, 109, 184

Advanced Very High Resolution Radiometer (AVHRR), 47, 116

agent-based modeling. *See* model: agent-based

Agricultural Model Intercomparison and Improvement Project (AgMIP), 146

agriculture, 37, 50, 72, 146, 186, 191

American Statistical Association, 154

animal, 40, 57, 68, 72, 100, 168, 169, 186, 204

Anthropocene, 4

antisubordination, 165

API. *See* application program interface (API)

application program interface (API), 89, 100, 111, 115, 209

apps, 49, **54**, 57, 188

artificial intelligence, 17, 21, 67, 77, 79, 115, 137, 185, 186; connectionist approaches, 79, 80, 83; GeoAI, 112, 115; hybrid approaches, 80; symbolic approaches, 78

Association of Data Scientists, 154

atmosphere, 18, 39, 52, 73, 74, 206, 208

Atmospheric Model Intercomparison Project (AMIP), 59

Atomicity, Consistency, Isolation, and Durability (ACID), 94

attribute data, 28, 30, 41, 48

attribute scale, 29

autocorrelation, 132

automated semantic derivation, 99

autonomous driving, 51

autonomous underwater vehicle, 38

autonomy, 21, 22, 165

Basically Available, Soft, Eventually (BASE), 94

BeiDou. *See* global navigation satellite system (GNSS)

bias. *See* big data: bias

big data, 2, 27, 40, 43; bias, 61, 66, 129, 163, 202; complexity, 13; exhaustivity, 13, 41, 71; extensionality, 13; extreme big data, 98; indexical, 12; origin of term, 7; relationality, 12; scalability, 13; speed and storage demands, 10; terminology, 6; value, 12; variability, 11; variety, 11, 71, 130; velocity, 10, 71; veracity, 11; visualization, 12; volume, 8, 71

big data for development (BD4D), 184

binary digit, 10

biodiversity, 15, 37, 67, 72, 176, 208

bot, 62

Brazil, 195

British Academy, 152

Cambridge Analytica, 164, 168

Canada, 180

capta, 7

captured data, 34

causation, 121, 127, 128, 151, 204

cell phone, 36, 48, 49, 56, 61, 98, 167, 184, 195

census data, 29, 31, 40, 64, 65, 69, 142, 155

China, 19, 47, 74, 116, 168, 181, 195, 207

Chinese Academy of Sciences, 2

Chinese Personal Information Protection Law, 181

class imbalance, 113

classification, 83

climate change, 64, 132, 140, 178, 181, 188, 189, 203

climate modeling. *See* model: climate

cloud computing, 95, 96, 100, 110, 116

CodaLab, 108

Code, Data, and Environment (CDE), 108

colocation analysis, 83

common task framework, 137

communication technology for development (ICTD), 183

compression, lossy, 98

computational social science, 148, 150

computational sustainability, 183, 190

consent, 168, 172, 173, 194

constructionism, 23, 199, 203

COOP. *See* Cooperative Observer Program (COOP)

Printed in the United States
by Baker & Taylor Publisher Services